TECHNICAL DOCUMENTATION and PROCESS

Jerry C. Whitaker
Robert K. Mancini

TECHNICAL DOCUMENTATION and PROCESS

CRC Press
Taylor & Francis Group
Boca Raton   London   New York

CRC Press is an imprint of the
Taylor & Francis Group, an **informa** business

CRC Press
Taylor & Francis Group
6000 Broken Sound Parkway NW, Suite 300
Boca Raton, FL 33487-2742

© 2013 by Taylor & Francis Group, LLC
CRC Press is an imprint of Taylor & Francis Group, an Informa business

Printed in the United States of America on acid-free paper
Version Date: 20120822

International Standard Book Number: 978-1-4398-6159-2 (Paperback)

Library of Congress Cataloging-in-Publication Data

Whitaker, Jerry C.
   Technical documentation and process / Jerry C. Whitaker, Robert K. Mancini.
      p. cm.
   Includes bibliographical references and index.
   ISBN 978-1-4398-6159-2 (pbk.)
   1. Technology--Documentation. 2. Communication in organizations. I. Mancini, Robert K. II. Title.

T10.5.W478 2012
808.06'66--dc23

2012028554

Visit the Taylor & Francis Web site at
http://www.taylorandfrancis.com

and the CRC Press Web site at
http://www.crcpress.com

This book is dedicated to my wife, Laura, who sets the bar

for people-management in the corporate world. (JCW)

I dedicate this book to my wife, Barbara, to our children, and

to our grandchildren. Your support and encouragement were

appreciated during the writing of this book. (RKM)

Contents

Preface

Our goal is to assist technical managers in developing appropriate process steps and documentation for effective and successful projects and products. Although not all guidelines are appropriate for all situations, the guidance given should help you develop process and documentation tools that address the particular needs of your organization.

A carefully defined strategy for developing a project or product is important whether on a large or small scale. A lack of planning can lead to misunderstanding, delay, and unhappy customers (and bosses). If you're on a tight schedule, that can translate into slipping delivery dates and penalty fees.

While there is no secret to developing effective process and documentation, it does require time, effort, and resources. And while these steps come with a price, the cost of minimal planning (or even no planning) can be high as well.

This book draws on the resources of several books previously published by CRC Press (and coauthor Jerry Whitaker) and many years of experience to provide comprehensive coverage of the subject matter. The level of detail provided in each chapter is intended to cover the major points and give readers sufficient information to develop their own procedures and tools. While impractical to provide ready-made solutions for every situation, we give guidelines, suggestions, and examples so that readers can develop their own specific plans. Considerable attention is given in the text to documentation and development of a style guide, the goal being to make the process of generating documentation easier for all of those involved. A selection of document templates is also provided to serve as a starting point for readers to develop their own templates.

The book is intended to help readers discover what they need to know in the area of process and documentation. Some readers no doubt will want additional information on one or more subjects covered in the book. There are a number of excellent texts available, many of which are referenced in the following chapters.

Although every effort has been made to cover the subject matter comprehensively, we also recognize that a relatively short, to-the-point book is probably most useful. The book, like any other, is a snapshot in time reflecting the experience of the authors.

For additional information on technical documentation and process, there are a number of interesting books in this field that may be of interest to readers. Offerings by CRC Press include the following:

- *Technical Writing: A Practical Guide for Engineers and Scientists* by Phillip A. Laplante. This book complements the traditional writer's reference manuals and other books on technical writing to provide real-world examples of technical writing. It also explores the various avenues for publishing your work and explains how to write for blogs, social networks, and other e-media.

- *Systems Engineering Focus on Business Architecture: Models, Methods, and Applications* by Sandra L. Furterer. This book is a straightforward manual for creating the Strategic Business Process Architecture in any organization. The author discusses how to consistently create high-quality business architecture models and ensure alignment between the business strategies and the change initiatives.

- *Systems Engineering and Architecting: Creating Formal Requirements* by Larry Bellagamba. This book presents formal requirements to help readers accomplish key systems engineering and architecting activities more efficiently. The formal requirements—explicit, executable, verifiable instructions—explain how to model systems behavior, make decisions, establish natural language requirements, and improve your systems engineering and architecting processes.

- *Managing Organizational Knowledge: 3rd Generation Knowledge Management and Beyond* by Charles A. Tryon, Jr. This book provides a clear, repeatable strategy for capturing organizational knowledge. The book presents innovative processes to help readers capture vital organizational knowledge.

It is our sincere hope that this book helps you define the process, document the plan, and manage your project. Wishing you great success!

Jerry C. Whitaker
Robert K. Mancini
Morgan Hill, California

About the Authors

Jerry C. Whitaker is vice president of Standards Development, Advanced Television Systems Committee (ATSC). He supports the work of the various ATSC technology and planning committees and assists in the development of ATSC standards, recommended practices, and related documents. ATSC is an international, nonprofit organization that develops voluntary standards for digital television.

A fellow of the Society of Broadcast Engineers and an SBE-certified professional broadcast engineer, he is also a fellow of the Society of Motion Picture and Television Engineers. Whitaker has been involved in various aspects of the electronics industry for over 30 years. His current CRC book titles include the following:

- *The Electronics Handbook*, 2nd edition
- *Electronic System Maintenance Handbook*, 2nd edition
- *AC Power Systems Handbook*, 3rd edition
- *The RF Transmission Systems Handbook*
- *Power Vacuum Tubes Handbook*, 3rd edition

Whitaker has lectured extensively on the topics of electronic systems design, installation, and maintenance. He is the former editorial director and associate publisher of *Broadcast Engineering* and *Video Systems* magazines, as well as a former radio station chief engineer and television news producer. Whitaker has twice received a Jesse H. Neal Award Certificate of Merit from the Association of Business Publishers for editorial excellence, and has also been recognized as Educator of the Year by the Society of Broadcast Engineers.

Robert K. Mancini is president of Mancini Enterprises, LLC, a business consulting, estate planning, and educational services corporation based in California. He has been involved in programming, product management, technical writing, sales/product training, and marketing in the software and aerospace industries for over 30 years.

His authored projects include the following titles:

- Image Focus manuals, marketing, and training materials, NewEra Software
- DSSI product marketing materials, Allen Systems Group

- PRO/JCL manuals, marketing, and training materials, Diversified Software
- DOCU/TEXT manuals, marketing, and training materials, Diversified Software
- JOB/SCAN manuals, marketing, and training materials, Diversified Software
- INFO/X manuals, marketing, and training materials, Diversified Software
- *How to Leave a Rich Legacy: The Personal Side of Estate Planning*
- *Azerbaijan: Artwork and History of Ancient Albania* (Editor-in-chief)

Mancini led the software documentation team at Diversified Software, and held a key role in responding to RFPs at Lockheed (which included documentation of systems design, prototype, development, installation, and maintenance). He is a frequent guest lecturer at Biola University and current member of the Professional Fiduciary Association of California (PFAC). While managing the documentation team at Diversified Software, Mancini was a member of the Society for Technical Communication (STC) and the American Society for Training & Development (ASTD).

1

Overview

1.1 Introduction

The subject area encompassed by "documentation and process" is broad in scope. It includes, but is not limited to, writing, organization, people management, project management, and problem-solving. Within each of these broad groups, additional distinctions can be identified; indeed, entire books have been written on these subjects. Beyond a deep dive into these and other topic areas, there is a need to see the big picture and integrate separate disciplines into a cohesive program or process.

The lessons of documentation and process can be applied across a wide variety of project types and organizational structures. This being the case, guidelines, suggestions, rules, and all the other elements that make up a process rarely are applied strictly to an organizational structure. Instead, that structure is executed and maintained by individuals. It is important, therefore, for those individuals to understand the process and take ownership of it.

While an ad hoc approach to a given project can be successful, the likelihood of success may be reduced due to the informal nature of the activity. Equally important, lessons learned during execution of one project may be lost and forgotten when the project is completed. One benefit of defining a process is that it compels the leadership to look at the big picture and to try to anticipate unforeseen challenges.

Process development is an inexact science. However, by documenting the lessons learned on past projects, the knowledge base of the organization grows. This learn-by-doing approach implies that a process developed for a given project may change over time as unexpected problems are encountered and solved.

Documentation skills come into play here as well. Documentation also, of course, touches many areas of business and plays a major role in the success of a project or product.

1.2 Plan for Success

Well-thought-out and documented plans increase the likelihood of a successful project. A structural process that cannot be implemented in a given organization is certain to fail, and the individuals tasked with carrying it out may fail as well. Success has many fathers; failure is an orphan. This well-known truism has been proved right countless of times in any number of organizations over a long period of time.

While the reasons for failure vary from one situation the next, certain common threads tend to emerge, including the following:

- **The goals were set too high.** In business, as in most everything else, you can't always get what you want. The needs of the organization must be balanced with the realities of resources, time, and capabilities. It is always a good practice to challenge individuals and organizations to produce their best; however, setting goals so high they are generally believed to be unachievable often results in team members giving up when the impossibility of the task ahead becomes clear.

- **The goals were set too low.** If a project is completed but the end result does not meet the need, then the effort can result in failure, or at least lost time as the project is rescoped and restarted. Individuals like knowing they are a part of something big, something important to the organization. A small project with only a minimal chance of having a positive impact invites lackluster participation and effort on the part of contributors.

Between these two extremes, naturally, there is a sweet spot where the organization and individuals within it are challenged with achievable goals and given the resources necessary to accomplish the task at hand. Although there are numerous factors involved, elements of successful projects may be generally summarized as follows and illustrated in Figure 1.1:

- **Adequate resources.** A project starved of resources is in trouble from the start. Typically, such resources translate into available personnel and money. The two are usually interrelated, of course. Other types of resource limitations include insufficient time made available on specialized machines or in research labs for focused work, restrictions on travel, and so on. The resources allocated to a project say something about the importance of the project to those tasked with carrying it out. If management doesn't think a project is important, then the employees are unlikely to put much effort into it.

- **Adequate time allocated.** The time needed to complete a project is closely related to the resources applied to complete the project.

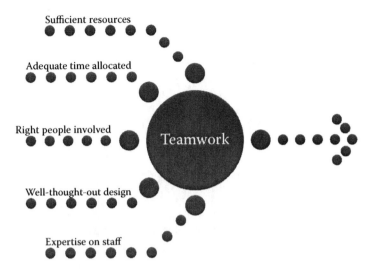

FIGURE 1.1
Elements of a successful project.

Adding resources usually shortens time, while reducing resources usually lengthens time. In some situations, the timeline is fixed in that a project (or product) needs to be completed by a specified date (e.g., promised ship date to the user). In other situations, the timeline is flexible, with no firm end date (although there is often a goal). As a practical matter, projects with no end date tend not to end. An end date is important—but as before with goals, the date needs to be realistic. Personnel working on the project need to understand what the deadlines are and why they are important to the success of the effort.

- **The right people in the room.** This challenge can be a tough one to solve. It is related to resources, but has a unique dimension as well. In any organization, there are key individuals who either direct development or make decisions about development. As such, it is important for them to be involved in the projects and processes that will fulfill the goals of the organization. Getting a slice of their time may be a challenge. If a key person is not involved, the project will tend to move on but with the risk that when the decision-maker becomes involved, it may be so late in the process that it derails or seriously delays the effort. One approach that often works well is to identify key points in the process where review takes place. This permits the key players to focus on other tasks most of the time, but step into a project at predetermined points to provide input or to suggest changes.

- **Well-thought-out design.** An organizational structure can have considerable inertia. Once a project has begun, it usually rolls forward. Occasionally, after some work in a particular direction, it may become clear to decision-makers that the original concept or technology was flawed, and the best approach is to go in an entirely new direction. This can be wrenching to individuals who have put months or even years into a project only to have it stuck on the shelf. The decision to kill a project or take a radically new approach is difficult to make for a number of reasons, not the least of which is the knowledge of wasted resources. Still, cutting losses may be the best approach, as the only thing worse than stopping a project mid-stream is to complete it, only to find that it is not what the customer wants or otherwise fails in the market.

- **Realistic view of expertise.** Some concepts are exciting and hold considerable promise in the marketplace. The problem is that they can also be very difficult to implement and may require expertise that just doesn't exist within the organization. As noted with goals, stretching the capabilities of an organization is generally good, but being unrealistic about what can be accomplished is not. There are a number of possible solutions to such a challenge. One would be to partner with another company or organization that has the needed expertise, with each party bring something different to the table. Business issues naturally come into play here; still, it may be an area worth exploring.

- **Building the team.** Teamwork is essential to a successful project—and one element of a successful project is the energy and enthusiasm behind it. A lack of energy and interest in a project is often the result of a combination of the challenges previously outlined. Most organizations can identify one or more projects that have plodded along to completion, but that generated very little energy, interest, or enthusiasm on the part of contributors. The end results are, predictably, uninspired as well. One important task of management in any organization is to motivate individuals. When they lack motivation, people do not produce their best work.

Just as individuals can be set up to succeed, or not, projects and processes can also be set up to succeed or fail. The role of management is to foster the former while preventing the latter.

Timing is another component in the success of a project. Some elements of timing are within the control of the organization and are set in the project timeline, as discussed previously. Other elements of timing are outside the control of the organization. For most things, there is a window of opportunity. The window may be large or small, and it can be very difficult to predict opportunities that are one, two, or three years out. There are numerous examples of a product that was offered to the market ahead of its time, when the market was not ready for it. There are probably many more examples where a product hit the market too late, having been overtaken by other technologies or approaches.

Market prediction can be a very difficult task. It involves—but is by no means limited to—market research, trend analysis, customer interaction, technology assessment, and luck.

1.3 Elements of Process

Process, within the scope of this book, focuses on the steps and structures needed to accomplish a set of stated goals. These steps include developing an organizational structure, coordinating the activities of participants, monitoring progress, documenting results, planning for unforeseen problems, and reporting the results of the work. The process developed for one group within a particular organization is often transferable to another group working on a related (or even unrelated) project. Such repurposing of management structures is helpful in that it reduces the time needed to begin work and tends to refine the individual process steps. Improvements in the process can be identified through documentation of things that worked well, and documentation of things that did not work as intended.

Process involves looking at the big picture and identifying the key steps necessary to get from here to there. The best structure is often a loose one, where guidelines and guideposts are established at key points along the way, but not so much detail and structure that it inhibits progress and creativity in the face of unforeseen events.

Invariably, documentation comes into play at all steps in a given process. Communication of ideas, problems, and solutions is essential to keep all members of the team on the same page, and top management advised of the status of important projects. One of the tools for the documentation specialist is a style guide that helps give structure to the overall effort and helps to maintain consistency and quality among documents from different groups.

Likewise, meetings are a critical element in any process. Meetings can serve as an opportunity to develop new ideas and concepts. They also give contributors a common vision of the task at hand. And, critically, they serve as a vehicle to make key decisions.

It is easy to find examples of process gone wrong. Meetings can turn into shouting matches. Documents can be inaccurate and difficult to read. The output of a long effort can fail to achieve the stated goals or requirements of the user. Such experiences reinforce the need to develop and refine process steps to maximize the probability of success. No process is perfect or can guarantee a winning product. In the end, success requires the right people working toward a common goal with the backing of management to provide the necessary resources to get the job done.

1.3.1 Documentation

It is difficult to thrive in the business world today without effective communications skills. Foremost among these is the ability to clearly communicate ideas in written form. As e-mail steadily replaces the telephone as the primary business communications tool, good writing skills have never been more important. Writing impacts all facets of business today.

A style guide encourages consistency among documents produced by different persons in different organizations. This leads to a cohesive internal and external image of the organization.

Development of a comprehensive style guide can be a major undertaking. Getting the buy-in of various decision-makers on the final details may be another project in itself. If done correctly, the end product will be worth the effort.

Companies looking to break out of a "me too" look sometimes take dramatic approaches to graphics and text in products and supporting literature. Sometimes this approach is very effective; other times it is simply a distraction. Most organizations develop their first style guide based largely on what has been done in the past, albeit informally. The current look and feel is usually a good place to start; at least it got the organization to where it is today.

A style guide may be developed internally, or by an outside consultant or firm. Both approaches have their benefits. Developing a style guide internally is almost always less expensive (even accounting for staff time), and there is no doubt that internal staff knows the product line and company better than any consultant. That said, an outside view can provide a valuable perspective, particularly if the company wants to make a break with the past. The best of both worlds would probably be some combination of inside and outside collaboration on the style guide.

For a style guide to be effective and useful, it needs to be applied over a long period of time. It makes little sense to spend a considerable amount of time (and money) developing a style guide only to have it changed a year later. For this reason, it is important to get buy-in from all stakeholders and to make them understand that whatever the style guide finally looks like, the organization will use it for years to come.

1.3.2 Social Media

At the time of this writing, organizations of all types have rushed to embrace social media as a way of gathering feedback on any number of subjects, and for marketing products in innovative ways to new customers (e.g., blogs and Facebook). While it is clear that social media is playing an important role in personal interaction, it is unclear how business organizations can effectively use it. Many efforts have been made; some have been quite successful—others not so much. The business aspects of social media deserve additional study. It may play an important role in one project, but be of little value in another.

Various new media resources (e.g., LinkedIn.com and Monster.com) exist for finding qualified job candidates and contractors. The task of identifying and recruiting talent is a discipline unto itself.

1.4 Putting It All Together

Developing effective documentation and a process that gets things done can be a complex undertaking. If done correctly, however, the benefits will be readily apparent in the output of the organization. Important elements, shown in Figure 1.2, include the following:

- **Documentation strategies.** There are a number of approaches to documenting a project, product, or facility. The approach used may be simple or complex, determined by the desired end result and the end user.
- **Developing a style guide.** The style guide serves as the foundation on which documents of various types are built. A well-designed style guide will speed the development effort and result in more effective communication of ideas and guidance.
- **Meetings.** Very little in business today can be done in a vacuum. Meetings are a crucial tool in bringing together ideas and stakeholders to move a project forward.

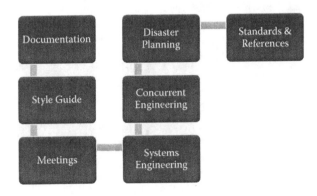

FIGURE 1.2
Elements of documentation and process examined in this book.

- **Systems engineering.** Small projects usually require little advance planning. Large projects, however, demand considerable planning, and a defined process for dealing with the unexpected.

- **Concurrent engineering.** Just as business does not operate in a vacuum, the methods for developing technologies and products have changed as well. The serial method of one step at a time is well understood and effective, but it can leave an organization at a disadvantage in a highly competitive environment. Concurrent engineering places key serial steps in parallel for increased efficiency.

- **Disaster planning and recovery.** A well-defined process for a company, project, or product always includes contingencies for when things go wrong. However, what happens when things really go wrong—at the scale of a natural disaster? This possibility too must be considered at an organizational level.

- **Standards and references.** A highly technical environment requires agreement on certain standards and reference points. Standardization of components, procedures, and protocols fosters interchangeability from one vendor to the next, usually at reduced cost to the consumer.

Each of these subject areas is examined in the chapters that follow.

One more thought about standards. Many companies and organizations are involved in standardization efforts that impact their particular interests and industries. Companies involved in the standards-setting process are able to see emerging technologies at an early point and may be able to shape their development to provide for enhanced utility and functionality. Companies not involved in the standardization process often have a variety of ready excuses for not participating (takes too much time, let somebody else figure it out, process is too slow, don't have anyone on staff available to contribute, etc.). Experience has shown, however, that companies involved in the standard-setting process are in the best position to capitalize on their own capabilities, and move the broader industry ahead as well. In business, it is always better to lead than to follow.

2

Equipment Documentation Strategies*

2.1 Introduction

Little in the technical professions is more important, exacting, or demanding than concise documentation of electronic physical plants, systems, and equipment. Yet this essential task, involving as it does both left and right brain activities—a combination of science and art—is all too often characterized as an adjunct skill best left to writers and other "specialized" talents. The predictable result is poor, incomplete, or incorrect documentation. The need for documentation is often underestimated because future modifications to systems, processes, and technology are often underestimated.

Neglecting the task of documentation will result, over time, in a technical facility where (for example) it is more economical and efficient to gut the existing wiring and start over rather than attempt to regain control of the documentation. Retroactive documentation is physically difficult and tedious, and seldom generates the level of commitment required to be entirely successful and accurate.

Inadequate documentation is a major contributor to the high cost of systems maintenance and for the resulting widespread distaste for documentation task; in that sense, bad documentation begets worse documentation.

Yet documentation is a management function every bit as much as project design, budgeting, planning, and quality control. Documentation is often the difference between an efficient and reliable facility and a misadventure. If the system designer does not feel qualified to attempt documentation of a project, that engineer must at the very least oversee and approve of the documentation developed by others.

The amount of time required for documentation can vary from perhaps 10 to 50 percent of the time actually required for the physical installation of a facility's equipment. Because this is often unseen work, few owners or managers understand its value, and because many engineers and technicians disdain paperwork, documentation often receives a low priority. In extreme

---

* Portions of the chapter are based on Baumgartner, F. and T.M. Baun, "Engineering documentation," in *The Electronics Handbook*, 2nd ed., Jerry C. Whitaker (Ed.), CRC Press, Boca Raton, FL, 2005.

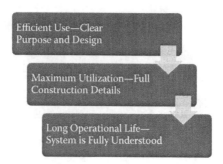

FIGURE 2.1
Benefits of a well-documented project.

cases, the technical staff may even see "keeping it in my head" as a form of job security, although that threat has little cachet with today's bottom-line oriented business managers.

Of course, the frustrating thing about documentation is that it is usually thought of as an accessory or afterthought to the act of construction, when in truth it is the very thing that allows the builder to best assure the long-term success of the construction.

As illustrated in Figure 2.1, a well-documented project pays dividends in three primary areas:

- Encourages efficient use of the project by providing a clear explanation of its purpose and design. Many projects are rebuilt or retired because of supposed obsolescence, when in fact the designers and builders anticipated future requirements and prepared the system to accept them. This can be lost without good documentation.

- Encourages maximum utilization of the project by providing a clear explanation of its parts and construction details. Future modifications can be made with full knowledge of the limits that must be respected as the system is fine-tuned throughout its operational life.

- Permits a longer effective operational life for the project as changes in technology or task may require. A product or facility that is poorly documented has little or no chance of being expanded to incorporate future changes in technology, simply because the task of writing documentation for the existing system is considered more onerous than reconstruction of the entire enterprise.

There is a psychological element at work here as well. Few people enjoy the task of mapping others' journeys—but we remain supremely confident that our trail has been so well blazed that anyone can follow in our footsteps. And it does not help matters that documentation is often seen as an embellishment rather than an element of construction in and of itself.

Conventional wisdom asserts that engineering talent and writing talent are often antithetical, providing yet another disincentive for many engineers to attempt to provide proper documentation. It is interesting to note that the "scientific method" requires that experimenters and builders be documenters; yet outside the laboratory, those actually engaged in project construction utilizing that scientific data may not be similarly tasked.

Ideally, documentation begins the moment a facility is first laid out. Hopefully, the time, tools, and other resources needed to fully document a facility are available from the beginning of the project. For the reasons mentioned previously, the value documentation returns will be many times its cost.

A properly documented physical plant or product exhibits two other major advantages. First, it is maintainable. Without documentation, a facility may require excessive time to repair or reconfigure. Second, the plant will be efficiently expandable. Without complete documentation, the system may expand in a haphazard manner with much of the full potential of some equipment unrealized.

2.2 Documentation Tools

Common office software offers the capability to easily update, reorganize, reprint, and analyze documentation and related information. There are numerous programs available specifically for document creation. These programs may fit a specific facility's needs or may leave something to be desired. In general, word processors, spreadsheets, CAD, and database programs can supplement or be used in lieu of an application-specific documentation program. Ideally, these programs allow data transfers between programs and permit a simple single entry to be recorded in several document files.

The keeper of the documentation will normally print out updated drawings and text as needed. Separate documents for each technical area are possible in this environment. In some large facilities, this individual documentation may be stored in each room where the documentation is required. Racks for architectural-type drawings and booklets are convenient for wall mounting.

2.2.1 Types of Documentation

There are three primary methods of system documentation: self-documentation, database documentation, and graphics documentation. In most cases, a mixture of all three is necessary, as illustrated in Figure 2.2. In addition to documenting the physical plant and its interconnections, each piece of equipment—whether commercially produced or custom-made—must be documented in an organized manner. Likewise, there are a number of aids for documenting the labeling of cables, wiring, and equipment.

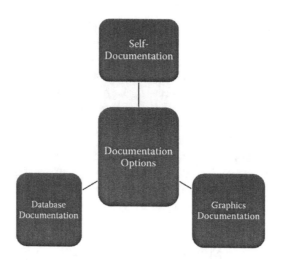

FIGURE 2.2
Relationship of documentation options.

2.2.1.1 Self-Documentation

In situations where the facility is small and very routine, self-documentation is possible. Self-documentation relies on a set of standard practices that are repeated. Conventional (analog) telephone installations, to those not familiar, appear as a mass of perplexing wires. In reality, the same simple circuits are repeated in an organized and universal manner. To the initiated, any telephone installation that follows the rules is easy to understand and repair or modify, no matter where or how large. Telephone installations are largely self-documenting. Familiarity with telephone installations is particularly useful, because the telephone system was the first massive electronics installation. It is the telephone system that gave us "relay" racks, grounding plans, demarcation, and virtually all of the other concepts that are part of today's electronic control or communications facility.

The organization, color codes, terminology, and layout of telephone systems are recorded in minute detail. Once a technician is familiar with the rules of telephone installations, drawings and written documentation are rarely required for routine expansion and repair. The same is true of many parts of other facilities. Certainly, much of the wiring in any given rack of equipment can be self-documenting. For example, a local area network (LAN) distribution hub will typically include a patch panel with destination points identified and one or more routers or switches. The wiring between each of these pieces of equipment is clearly visible, with all wires short, and their purpose obvious to any technician familiar with the rules of LAN interconnection. Color-coded patch cables further simplifies the process. As such, further documentation is largely unnecessary.

The key to self-documentation is a consistent set of rules that are either obvious or themselves clearly chronicled, and technicians who are familiar with those rules. If the rules are broken, self-documentation will quickly break down.

By convention, there are rules of grounding, power, and signal flow in all engineering facilities. In general, it can be assumed that in most communications facilities, the ground will be a star system, the power will be individual 20 A feeds to each rack or room, and the signal flow will be from top to bottom.

Rules that might vary from facility to facility include color coding, connector pin-outs, and rules for shield and return grounding.

To be self-documenting, the rules must be determined and all of the technicians working at the facility must know and follow the conventions. The larger the number of technicians, or the higher the rate of staff turnover, the more important it is to have a readily available document that clearly covers the conventions in use.

A facility that does not have written documentation is not automatically self-documenting; quite the contrary. A written set of conventions and unfaltering adherence to them are the trademarks of a self-documenting facility. Even a single sheet of paper documenting connector pin-outs, color codes, and cable numbers might suffice, and would certainly be the first step toward self-documentation.

While it is good engineering practice to design all facilities to be as self-documenting as possible, there are limits to the power of self-documentation. In the practical world, self-documentation can greatly reduce the amount of written documentation required, but seldom can replace it entirely.

2.2.1.2 Database Documentation

As systems expand in size and complexity, a given set of conventions no longer answers all of the questions. At some point, a wire leaves an equipment rack, and its destination is no longer obvious. Likewise, a unique configuration of equipment will often require written documentation.

Database documentation records the locations of both ends of a given circuit. For this, each cable must be identified individually. There are two common systems for numbering cables: ascension numbers and from-to-coding. In ascension numbering schemes, each wire or cable is numbered in increasing order (1, 2, 3, ..., etc.). In from-to-coding, the number on each cable represents the source location, the destination location, and normally some identification as to purpose and a unique identifier. For example, a cable labeled 31-35-E6 might indicate that a cable went from a piece of equipment in rack 31 to another unit in rack 35 and carries Ethernet traffic; it is also the sixth cable to follow the same route and carry the same class of signal.

Each method has its benefits. Ascension numbering is easier to assign, and commonly available preprinted wire labels can be used. On the downside, ascension numbers contain no informational hints as to purpose or path. Ascension numbers can have added data, for example, a purpose code.

From-to-codes can contain a great deal more information without relying on printed documentation records. This makes repair and emergency modification rapid and elementary.

Whatever numbering system is used, a complete listing must be kept in a database of some kind. In smaller installations, this is simply a book that contains a complete list of all cables, their source, destination, any demarcations, signal parameters, and the like.

The ability to organize and reorganize this information is important. Typically, the documentation is organized by location and signal type. Having a separate printout for each location (rack) that lists all of the wiring, and where it goes, is important. Each cable has two ends, so the same wire would be included in each of the two location's printout. Likewise, being able to print out all of the locations of a given signal type can be useful.

2.2.1.3 Graphics Documentation

Electronics is largely a graphical language. Schematics and flowcharts are more understandable than net lists or cable interconnection lists. Drawings are highly useful to gain overall system knowledge. Normally, these drawings are created to display various sections or locations in a facility. Typically, the interconnections displayed are denoted with the same numbers that appear in the database documentation.

2.2.2 Labeling

There are many ways to label cables and wires. The simplest and cheapest is the wire-tie tag label that can be written upon with a pen. Tags using wrap-around clear plastic protectors can be printed, making them more legible, durable, and less likely to tear off when pulled. Printed tags often take longer to produce; however, with the right software they can be generated as part of a documentation program. Hand-held labeling systems are available that allow creation of professionally printed labels at the job site. These devices allow an installer to create the cable labels as the connections themselves are being made, as well as equipment designation labels and in-house asset number tags.

The high end of cable labeling is the use of stamped-in or cable printing. Typically, this method applies a printed marking either with ink or by slightly melting (branding) the cable jacket every foot or so. The benefit is that the cable can be identified anywhere along its path. The costs and time involved make this method practical for only very large facilities.

Typically, each end of the cable has the same label identification. In addition, machine connection information is highly desirable. If a machine is removed for service, it is very useful to have the cables labeled in such a manner that it is not necessary to check the documentation to replace it. Too much information seldom presents a problem.

2.2.3 Other Documentation Tasks

Plant documentation does not end when all of the circuit paths in the facility are defined. Each circuit begins and ends at a piece of equipment that can be modified and reconfigured, or fail and require repair. Keep in mind that unless the lead technician lives forever, never changes jobs and never takes a day off, someone less familiar with the equipment will eventually be asked to return it to service. For this reason, a documentation file for each piece of equipment should be maintained.

Equipment documentation should contain these key elements: the equipment manual, the modification record, the configuration information, and ideally, the maintenance record.

The equipment manual is the manufacturer's original documentation. This is often shipped with the equipment, or sold as a separate item (verify that a copy of the documentation will come with each piece of equipment purchased; budget as necessary). If there are a number of pieces of identical equipment, the facility may need only one copy of each manual (unless equipment transfer to sale is anticipated in the future). The manuals must be organized in such a manner that they can be easily located. The most common organization is to set aside an area with shelves and or filing cabinets and arrange the manuals alphabetically by manufacturer name. Manuals are often physically larger than the equipment they document. Also, manuals come in very different formats and sizes. When a facility is first laid out, adequate space that is secure but available to technicians must be set aside. Typically, the manuals are kept at the site where the equipment is installed if practical.

Many pieces of equipment, over time, will require updates or modification. This may be at the manufacturer's suggestion, due to changing use, or in time, modifications simply to keep the equipment operating. Such changes should be recorded in the documentation for that equipment.

Equipment documentation is increasingly being supplied electronically, as Acrobat PDF files or a similar commonly available file format. For certain equipment, the sheer volume of documentation that may be supplied makes it very difficult to practically store as printed books. The benefits of electronic documents, of course, are that they can be stored in a central file server and made available to engineers and technicians wherever they may be working. Electronic folders can be created that include not only the equipment documentation, but also maintenance notes, firmware update files, common configurations, and license keys. These documentation files should be treated like any other production file.

Printed documentation is still appropriate, however, for devices such as servers on which the electronic manuals are kept. It makes little sense to store the documentation to a particular server on that server. Should the server fail, then the documentation for that unit would be unavailable.

2.2.3.1 Operator/User Documentation

User documentation provides, at its most basic level, instructions on how to use a system. While most equipment manufacturers provide reasonably good instruction and operations manuals for their products, when those products are integrated into a system, another level of documentation may be required. Complex equipment may require interface components that need to be configured from time to time, or various devices may be incompatible in certain modes. For many systems, such information resides in the "heads" of certain key users and is passed along by word of mouth or informal notes. This level of informality can be very risky, especially when changes take place in the user group or maintenance staff. The result of such a situation may be differing interpretations between operators and maintenance people regarding how the system normally operates, resulting in maintenance personnel spending considerable time tracking nonexistent errors and perhaps making mistakes along the way.

2.2.3.2 Schematic Documentation

Engineering documentation describes the practices and procedures used within the industry to specify a design and communicate the design requirements to technicians, users, and vendors. Documentation preparation should include, but not be limited to, the generation of technical system flow diagrams, material and parts lists, custom item fabrication drawings, and rack and console elevations. The required documents include the following:

- Signal flow diagram
- Equipment schedule
- Patch panel assignment schedule
- Rack elevation drawing
- Construction detail drawing
- Duct and conduit layout drawing
- Single-line electrical flow diagram
- Bill of materials (BOM)

2.2.4 Symbols

There is a wide variety of generally accepted industry standards for electronic component symbols to represent equipment and other elements in a system. Standard symbols should be used whenever possible. In special cases, however, it may be necessary to develop custom symbols for a particular project or product. Figure 2.3 shows some common component-level symbols currently used in electronics.

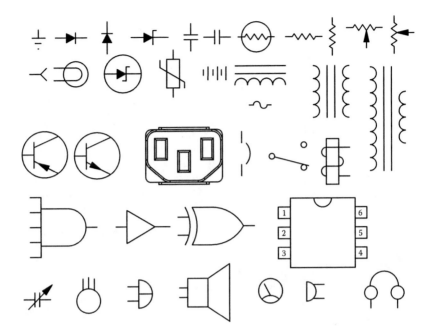

FIGURE 2.3
Common symbols for electronic circuits.

The proliferation of manufacturers and equipment types makes it impractical to develop a complete library, but, by following basic rules for symbol
design, new component symbols can be produced easily as they are added
to the system.

For small systems built with a few simple components, all of the input and
output signals can be included in one symbol. However, when the system
uses complex equipment with many inputs and outputs with different types
of signals, it may be necessary to draw different diagrams for each type of
signal. For this reason, each component requires a set of symbols, with a
separate symbol assigned for each signal type, showing its inputs and outputs only.

If abbreviations are used, be consistent from one drawing to the next, and
develop a dictionary of abbreviations for the drawing set. Include the dictionary with the documentation.

2.2.5 Cross-Referencing Documentation

In order to tie all of the documentation together and to enable fabricators,
installers, maintenance personnel, and end users to understand the relationships between the drawings, the documents should include reference
designations common to the project. In this way, items on one type of document can be located on another type. For example, the location of a piece of

equipment can be indicated at its symbol on the flow diagram so that the technician can quickly identify it on the rack elevation drawing and in the actual rack.

A flow diagram is used by the installation technician to assemble and wire the system components together. All necessary information must be included to avoid confusion and delays. When designing a symbol to represent a component in a flow diagram, include all of the necessary information to identify, locate, and wire that component into the system. The information should include the following:

- A generic description of the component and its abbreviation. Include it in the project manual reference section and in the notes on the drawing.
- When no abbreviation exists, create one. Make sure the abbreviation is clear and does not conflict with other facility-specific or industry-specific abbreviations.
- Manufacturer of the component.
- Model number of the component.
- All input and output connections with their respective names and numbers.

2.3 Specifications

Specifications are a compilation of knowledge about how something should be done. A systems engineer condenses years of personal experience, and that of others, into the specification. The more detailed the specification, the higher the probability that the job will be done right.

The systems engineer must provide support and guidance to contractors during the procurement, construction, installation, testing, and acceptance phases of a project. The systems engineer can assist in ordering equipment and can help coordinate the move to a new or renovated facility. This can be critical if a great deal of existing equipment is being relocated. In the case of new equipment, the system engineer's knowledge of prices, features, and delivery times is invaluable to the facility owner. The system engineer can thus

- Clarify details
- Clarify misunderstandings about the requirements
- Resolve problems that may arise
- Educate contractors about the requirements of the project
- Assure that the work conforms to the specifications

- Evaluate and approve change requests
- Provide technical support to the contractors when needed

Bibliography

Baumgarter, F., and T.M. Baun, "Engineering documentation," in *The Electronics Handbook*, ed. Jerry C. Whitaker, Boca Raton, FL, CRC Press, 1996.

Beizer, B., *Personal Computer Quality*, New York, Van Nostrand-Reinhold, 1986.

Martin, J., *Software Maintenance*, Englewood Cliffs, NJ, Prentice Hall, 1983.

Singer, L. M., *The Data Processing Manager's Survival Manual*, New York, Wiley, 1982.

Watts, F. B., *Engineering Documentation Control Handbook*, Urbana–Champaign, IL, University of Illinois, 1993.

Whittaker, J. C., E. DeSatnis, and C. Robert Paulson, *Interconnecting Electronic Systems*, Boca Raton, FL, CRC Press, 1992.

3

Developing a Style Guide

3.1 Introduction

A style guide is intended to assist the authors and editors of technical documents, reducing collaboration time by eliminating the need to combine differing formats into a single document. The guidelines given, typically along with a template, are intended to make the process of generating documents easier for all of those involved in the effort. Not all guidelines are appropriate for all situations; however, for the sake of consistency across all technical documents within an organization, the style guide should be followed to the extent possible.

3.1.1 Definition of Documents

The first step in developing a style guide is to identify the range of documents that the guide is intended to address. Document types can include the following:

- Technical marketing piece
- User manual
- Standards and practices
- Application note
- Technical bulletin
- Internal communications document

It is important to clearly define each document type, as that will guide the initial scope of the work. For example, an application note might be defined as a document that "states specifications or criteria that are not strictly necessary for effective implementation and interoperability, but that are thought to be advisable and may improve the efficiency of implementation or reduce the probability of implementation errors." A technical bulletin might be defined as "a document that incorporates information regarding proper operation of hardware or software systems." Whatever the definitions, they should be clearly stated in the style guide.

Documents also need to be written to a particular target audience, with an agreed-upon level of technical detail. Hardware and software manuals, for example, need to address requirements, installation, and operation. They should be written at a level that is understood by the target audience: enough information for the beginner, but not too much detail for the experienced user.

Whatever the document types identified by the company, the definitions should include a profile of the intended reader so writers and editors can fashion the text appropriately. It is often helpful to give an early draft of a document in development to one of the intended users (readers) and ask for feedback on how well the text and illustrations convey the subject matter. This step is best done late enough in the process so a reasonably well-refined version is available, but not so late that changes are difficult to make if needed.

3.2 General Structure of Documents

To maintain consistency across technical documents, a template should be created and applied in the creation of new documents or modification of old documents. Although each type of document, and indeed each subject matter, may require special treatment, the global outline should be followed to the extent practical. One approach is shown in Figure 3.1.

Consider the following template for a detailed technical document:

a. **Cover page**: Includes title of document, document number, date, and revision history (if appropriate).

b. **Organization page**: Includes a description of the organization or company. This is always page 2 of the document. This page may include a revision history table.

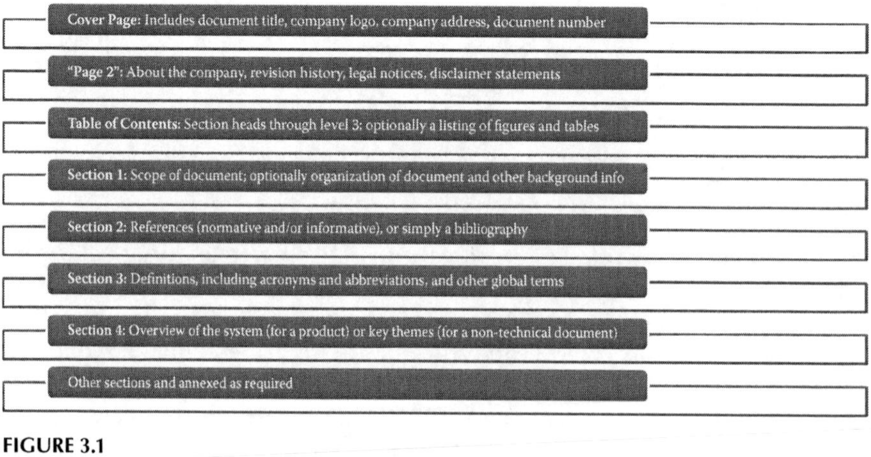

Cover Page: Includes document title, company logo, company address, document number

"Page 2": About the company, revision history, legal notices, disclaimer statements

Table of Contents: Section heads through level 3; optionally a listing of figures and tables

Section 1: Scope of document; optionally organization of document and other background info

Section 2: References (normative and/or informative), or simply a bibliography

Section 3: Definitions, including acronyms and abbreviations, and other global terms

Section 4: Overview of the system (for a product) or key themes (for a non-technical document)

Other sections and annexed as required

FIGURE 3.1
Key elements of an example technical document.

   c. **Table of contents**: Includes first-, second-, and third-level heads (e.g., 1, 1.1, 1.1.1), and other levels as appropriate.

   d. **Section 1—Scope**: Includes the purpose and organization of the documents as individual subheads (if needed).

   e. **Section 2—References**: Includes normative references (required) and informative references (as appropriate) as individual subheads.[*]

   f. **Section 3—Definitions**: Includes compliance notations, acronyms and abbreviations, global terms, syntax, and other definitions (as required) as individual subheads.

   g. **Section 4—Overview**: Provides an executive summary of the system or subject described in the document.

   h. **Other sections and annexes**: As required.

   i. **Running header**: Document number (left), document title (center), and date (right), with a medium weight (0.75 point) rule running the column width.

   j. **Page numbers**: Centered at the bottom of the page.

Different document types will require specific templates. The idea, however, is to provide a consistent structure to documents from a given organization or company.

3.2.1 Page Layout

The most important contribution of the document authors and editor is their expertise in explaining the concepts, procedures, and constraints of the subject matter. Attention to the details of page formatting, naturally, is of secondary importance. Still, providing general guidelines in a template at the beginning of a project can be helpful. An example is given below, and summarized in Figure 3.2.

   a. **Fonts**: In a technical document, the fewer fonts the better. For body text, use a 12 point serif face for best readability; Times New Roman in either its True Type or Open Type version is recommended. For headings, tables, and artwork callouts, use a sans serif face; Arial (True Type) or Helvetica (Open Type) are recommended. Arial or Helvetica can also be used within the body text to indicate commands and code words (9 point is recommended). For equations and

---

[*] A normative reference is a document that is required in order for a given concept to be understood. For example, the document might state "The security subsystem shall be as defined in Doc. #123." This approach eliminates the need to copy information from "Doc. #123" into the new text. It also eliminates the troublesome situation where the same specification is located in two or more documents. An informative reference, on the other hand, provides additional background information.

Fonts	
• Body text = Times New Roman, 12 pt., justified	
• Headings = Arial or Helvetica, ragged right	
• Level 1 = 12 pt. bold, Level 2 = 12 pt., Level 3 = 11 pt., Level 4 = 10 pt.	

Margins	
• Paper size = 8.5 by 11 inches	
• Margins = 1 inch top, bottom, sides	
• Set formatting for facing pages if printed, single pages if PDF distribution	

Numbering	
• Figure and table numbers use section number prefix (or annex as appropriate)	
• Page numbering sequential from beginning to end	
• Footnotes sequential from beginning to end	

Figures	
• Sans serif font (Arial or Helvetica), 10 pt. typical	
• Artwork floats on page; no box rule	
• Ideally sized for full page width (6.5 inches)	

Tables	
• Arial or Helvetica, 9 pt. bold for headings, 9 pt. for cells; all text flush left	
• Table cutline placed above the table	
• Ruling: top, bottom, and sides = 0.75 pt., inside rows and columns = 0.25 pt.	

FIGURE 3.2
Example style guide parameters.

Greek characters, use Symbol (either True Type or Open Type). For hyperlinks, use the base font of the paragraph tag with blue characters and blue underline. Avoid including hyperlinks to documents or websites that are likely to change over time.

b. **Margins:** Set the page margins for 8.5 by 11 inch paper and allow 1 inch margins on the top, bottom, and sides.

c. **Justification and indentation:** All body text should be justified with the first line indented. The first paragraph after a subhead, however, should not be indented. Bulleted or numbered lists are usually indented and set justified.

d. **Table/figure numbering:** The numbering of tables and figures in the document should follow the section numbers in which the elements appear. For example, the first table in Section 5 of a document would be Table 5.1. For tables and figures in an annex, include the annex letter followed sequentially by the number, for example, Figure A4.1.

e. **Tables:** Providing specific guidelines for table construction is difficult because of the wide range of information required to be presented in tabular form. Consider the following general guidelines:

- Use a medium weight (0.75 point) rule around the table.
- Use thin rules (0.25 point) within the table to define the cells, except for the heading row (or column), which should be separated from the contents by a 0.5 point line.

- Set the cell text left (using a sans serif font, as described previously).
- Use tabs (or indentation) as necessary to establish a tiered level of importance or to describe a specific structure of data or syntax.
- Place the table cutline above the table itself.
- Place the table as close as possible to the body text that refers to it, usually after a paragraph break. Avoid tables that break across pages.
- For cell text, use 9 point type, except in the case of semantic element names, which are usually 9 point bold type.
- For heading text, use 9 point bold type.

The following rules should be used with 'for' and 'if' statements when used with syntax in tables:

One space after 'for', 'if', semicolon, and between ')' and '{'

No space on either side of ' = ', ' = =', '<', or '>' except when next to syntax names or text

No space between <syntax> and open-close-parenthesis; e.g., 'descriptor()'

No space between empty parenthesis; e.g., '()'

One space between <syntax> and an opening bracket; e.g., "foo_descriptor() {'

Operators 'i', 'j', and 'k' should be lowercase when used in a table

f. **Hex numbering:** When a bit field is greater than 16 bits long, include a space after the hex characters representing the four sets of four bits in the first 16 bit hex field and after each subsequent set of 16 bits. This will break the presentation to improve readability; for example, "0x4454 4731". Note that appropriate text should be used to indicate a range of hex values; for example, "0x01" through "0x11".

g. **Schema namespace:** The path to schema available on a website should adhere to the following structure, adding additional elements if needed: http://www.domain.com/XMLSchemas/<application>/<year>/<version>/<schema-class>/....

h. **Figures:** As with tables, the range of figures required for a technical document is substantial. Consider the following general guidelines:

- Use a sans serif font (as described previously) and size the illustration to be readable at the size of the printed page (usually 8.5 by 11).
- For complex diagrams, consider placing the artwork at a 90° angle to provide for a larger finished illustration.
- Authors and editors are, of course, free to develop their drawings on whatever program they are comfortable with. Still, consider using a professional drawing package, such as Adobe Illustrator

or Microsoft Visio, to create drawings. Such programs offer features and time-saving elements not available in more general-purpose applications.

- Artwork should be embedded in the document. For imported images, EPS, JPG, PNG, and TIFF files are recommended.
- For professional printing, any line art should be scanned at 600 dpi or greater, preferably at the finished physical size.
- Black and white photographs should be scanned at 300 dpi or greater, again preferably at the finished size. Scanned black and white images should have a minimum highlight dot of 8 percent and a maximum shadow dot of 90 percent.
- If art consists of computer-screen captures, use a screen capture application capable of saving screen images at the proper resolution for printing.
- Place the artwork as close as is practical to the text that refers to it, usually after a paragraph break.
- Place the figure cutline below the artwork with the figure number boldface (i.e., "**Figure 5.1** Descriptive text here.") Let the artwork float centered (no box rule).

These examples are only suggestions to be considered as a starting point for developing a style guide that meets the needs of a given organization.

To facilitate interchangeability of documents, specify a recommended document-creating/editing program (e.g., Microsoft Word). Final preparation of the document for publication may be produced in another program (e.g., Adobe FrameMaker), but this will add to the complexity of the documentation project. Final output of the document may be done in one or more formats; Adobe Acrobat is one common format. For documents available electronically, it may be advisable to use a change-protected software package (or at least set security and permissions restrictions on the document to prevent unauthorized distribution or modifications).

3.2.2 Standard Writing Practices

Just as page layout is important to be defined within the style guide, so are basic writing practices such as punctuation, hyphenation, and capitalization. The *New York Public Library Writers Guide to Style Usage* (NYPL) is an excellent reference for basic writing practices. (See "Resources.")

3.2.2.1 Conciseness and Precision

Technical writing is characterized by two features: conciseness and precision (Laplante, 2011). These are qualities of the writing style and of the authors and editors of the work. Unfortunately, there is no set of universally agreed standards for good writing.

There are, however, some standards for specialized document types. For example, IEEE 830-1993 is a standard for specification documentation for "Requirements Engineering for Systems and Software" (IEEE, 1993). This is typical of technical writing used in many industries, and helps to provide a high-level description of the functions and features of a proposed system.

IEEE Standard 830 proposes eight desirable qualities for Software Requirements Specifications (SRS) documents. Five of these are relevant to any type of technical writing:

- **Correct**. The information presented is grammatically and technically correct.
- **Clear**. Each sentence, related groups of sentences, or related sections of the document can have only one interpretation. The text is unambiguous.
- **Complete**. There is no missing relevant or important information.
- **Consistent**. There are no contradictions within the document, and the document is in agreement with other associated documents.
- **Changeable**. The structure of the document is set up to handle changes, such as when errors or omissions are identified, or new information is discovered. Along the same lines, the document should be logically numbered, stored in a convenient electronic format, and compatible with common document processing and reading tools.

These guidelines should be applied to any document, although some document types are more critical than others. For example, an industry standard must be held to a higher level than an internal informational report.

3.2.2.2 Consistency in Sentence Structure

Use introductory sentences when they are able to provide more information than is contained in a heading or table title (Leitch, 2004). Introductory sentences should end with a period. When introducing a table or figure, do not end the introductory sentence with a colon.

Use the articles (a, an, the) in the body and cell body text, but not in headings. This will help with the readability in the sentence particularly with ESL (English as a Second Language) readers. For example, write "Indicates the alarm status" rather than "Indicates alarm status."

Table entries should not mix sentence fragments and full sentences within a cell or within a whole table.

List entries should be consistent. They should contain either all sentence fragments or all full sentences. This will help make usage of periods consistent since sentence fragments should never end with a period.

Within the body text, capitalization (aside from the standard "sentence case") should be used with caution. Many words are capitalized by authors when there is no need to do so. When a word has a particular special meaning different from the U.S. English meaning, that meaning should be stated, and the word or phrase capitalized when used. To call attention to a term or phrase, consider using italics instead.

3.2.2.3 Punctuation

Table 3.1 lists some of basic examples of how to use punctuation correctly and consistently.

3.2.2.4 Character Styles

Table 3.2 lists some basic examples of how to use character styles within the body text as well as within table cells.

Consider the following general guidelines regarding character styles:

- Apply italic formatting to the word "see" when introducing a cross-reference. This includes cross-references to specific manuals or sections within a manual, as well as nonspecific references. For example, apply italic formatting to the word "see" in both of these instances:
  - *See* the NEO MXA-3901H Installation and Operations Manual.
  - *See* your corresponding product manual.
- Only apply bold formatting to the menu names (and other items that would normally be bolded) in a table when they are used within a description. Do not apply bold formatting to them when they are stand-alone names in a separate column.

The purpose of applying bold formatting is to make something stand out to a user. If an item is already "standing out" in an identified column, then applying bold formatting becomes redundant and unnecessary. See the example in Table 3.3.

3.2.2.5 Headings and Titles

All headings should be set in "title case," except first-level heads, which are in all capitals. Keep titles short and to the point so that the headers are easy

TABLE 3.1

Recommended Punctuation

Punctuation	Style description
Period	Always use a period at the end of sentences that introduce figures or tables.
	Always place periods within final quotation marks (see Abrams 1994, 246).
Comma	Always include a comma before a conjunction in a list with more than two items.
	Never use commas at the end of a list. Use periods if it is a full sentence; otherwise, leave unpunctuated.
	Always place commas within final quotation marks (see Abrams 1994, 246).
Hyphen	Avoid leaving hyphens open-ended. Modify the Page Layout and Paragraph "Hyphenation" software options.
	Always specify "Smart Spaces" software option.
	Do not insert spaces on either side of a hyphen. Close the gap with a word on either side.
	Do not use a hyphen when an em-dash or en-dash should be used instead. Hyphens are used to join words. Dashes are used to join phrases or ideas. See "Dashes," below.
Dashes	Use an *em-dash* (—) in sentences instead of a comma, colon, or parenthesis to set off information (see Abrams 1994, 263).
	Use an *en-dash* (–) to represent the word "to" or "through" between two numbers or in expressions of time. It also marks the division between two elements, one or both of which already contains a hyphen (see Abrams 1994, 264). It also represents a negative or minus sign.
Colon	Use a colon only after a complete sentence.
	Do not use a colon at the end of sentences that introduce figures or tables.
Quotations	Use "smart" double quotes rather than straight quotes by specifying the "Smart Quotes" software option.
Slash (forward and backward)	Do not put spaces before or after slashes.
	Avoid slashes at the end of lines.
Less-than (">") sign (for navigation)	Use the less-than sign (">") to delineate menu and/or field names when documenting software navigation.
	Do not put spaces between the less-than sign (">") and the menus/ fields found on either side. For example, there are no spaces within this instruction:
	"Go to **Page Layout>Page Setup>Margins**."
	Do not bold the less-than sign (">") when used for navigational purposes. Do bold the menu/field names on either side.

Source:　Leitch, *Documentation Department Style Guide*, Edition A, Leitch Corporation, Toronto, Canada, 2004.

TABLE 3.2

Character Styles

Term or convention	Style description
Bold	Indicates dialog box, property sheet, field, button, check box, list box, combo box, menu, submenu, window, list, and selection name.
Italics	Indicates e-mail address, book, publication, first instance of a new term, and specialized words that need emphasis.
CAPS	Indicates a specific key on the keyboard (e.g., ENTER, TAB, CTRL, ALT, or DELETE) or a defined acronym.
Code (Courier New font)	Indicates variables or command-line entries, such as data typed into a field. May also be used for syntax.
>	Indicates the direction of navigation through a hierarchy of menus and windows.
<u>Hyperlink</u>	Indicates a jump to another location within the document, a different document, or a web address.
	Indicates important information that helps the reader to avoid and troubleshoot problems.

TABLE 3.3

Character Emphasis Example

Button	Style description
Save	Do not apply bold formatting to **Save** in the left column. The "Button" column title clearly indicates that the items beneath will all be buttons. The button names are easy to identify in a table format, so the application of bold formatting is unnecessary in this case.
	Do apply bold formatting to the button name **Save** in this "Style Description" column; otherwise, it will not stand out in the text.

to read when flipping through the document. For example, use "Chapter 1: Introduction" rather than "Chapter 1: Introducing the MXA-3901H."

Typical manual layouts include the following major section titles:

- Preface
- Contents
- Introduction
- Installation
- Operation
- Specifications
- Servicing Instructions

Consider using the gerund form for H1/H2/H3 headings. Making the heading end in "-ing" helps users get a sense of "doing" something. For example, use titles such as "Using," "Creating," "Connecting," "Understanding," and so on. Introduction and Specification chapters can be implemented in procedural chapters such as Installing and Operating. Typical headings for an Installation chapter might include the following headings:

- Unpacking the Module
- Preparing the Product for Installation
- Checking the Packing List
- Setting Jumpers
- Installing Modules
- Making Connections
- Removing Modules

Avoid using personal articles (e.g., "your," "you") in titles and headings. For example, "Connecting the MediaFile Dongle" rather then "Connecting your MediaFile Dongle."

As previously mentioned, headings are capitalized. When you have a hyphenated word in the heading, the general rule is capitalize the word after the hyphen unless it is a short, unimportant word such as "to." If you have a compound word that is hyphenated in a title, capitalize both words if they have equal weight or if the second word is a noun or proper adjective. For example, "Cross-Reference" or "Run-Time." Do not capitalize the second word if it is another part of speech or a participle modifying the first word. For example, "How-to" or "Take-off."

3.2.2.6 Bulleted and Numbered Lists

Listings in the body text can be numbered, unnumbered, or bulleted (Leitch, 2004). Use numbered lists for sequential steps (like a procedure, where users must do things in a certain order), and bulleted lists for items that are not sequential (items that do not necessarily need to be done in a specific order).

Punctuation should be consistent throughout a listing and follow proper grammar. The first word of each item should be capitalized. If a listing item is not a complete sentence, no punctuation is used. All items within a group should be complete sentences or incomplete sentences. A numbered list can be useful in a document, particularly if the text refers to specific items in the list.

Some general style guidelines surrounding bulleted and numbered lists include the following:

- If a complete sentence introduces a list, the sentence ends with a colon. If an incomplete sentence introduces a list, then do not punctuate the sentence.
- The preference is that lists should be introduced with a complete sentence.

- All bulleted points begin with a capital letter.
- Punctuation is not used in the list unless each bullet is a complete sentence, in which case each item ends with a period. Make all list items consistent.
- Avoid beginning a list item with a number. Numbers that begin a sentence in a list need to be spelled out, even if they are greater than 10.

3.2.3 Image File Formats

Some elaboration on image formats is appropriate. A vector file creates an image as a collection of lines rather than as a pattern of individual pixels (bit-mapped graphics). Vector files are much easier to edit than bit-mapped graphics (objects can be individually selected, sized, moved, and otherwise manipulated) and are preferred for professional illustration purposes. Because they are scale- and resolution-independent, vector images can be enlarged without loss of sharpness. Preferred vector file formats are listed below:

- Adobe Illustrator (.ai) is well-suited for creating high-quality professional graphics.
- Adobe Portable Document File (.pdf) is a file format that allows a document to be transferred to another type of computer system without losing the original formatting or font information.
- Encapsulated PostScript (.eps) format is a high-resolution graphic image stored in the PostScript language. The .eps format allows users to transfer high-resolution graphics images between applications. The images can also be sized without sacrificing quality.
- Microsoft Visio (.vsd) is well-suited for creating high-quality professional graphics, particularly when those illustrations are embedded in Microsoft Word files.
- AutoCAD is a popular computer-aided design program with a long history. Many features are available for documentation applications.

PostScript is a page description language (PDL) that is capable of describing the entire appearance of a formatted page, including layout, fonts, graphics, and scanned images. Because a PostScript file is device-independent, it can be printed on an imagesetter or any PostScript-compatible printer and will retain the original formatting. It does not provide compression, however, and so files can be quite large.

A *halftone* is a printed reproduction of a photograph (or an illustration other than line art). It uses evenly spaced dots of varying sizes to simulate shades of gray. Dense patterns of larger dots produce dark shades, and less dense patterns of smaller dots create lighter shades.

While the file formats listed above are the preferred types, almost any illustration that conveys the necessary information can typically be used by the author. To meet the requirements of publication, some illustrations may need to be reworked or redrawn by the publishing department or staff. This will add extra time to the documentation project, and should be discussed in the planning stages.

The use of color in illustrations and photographs can help the reader understand the subject matter. A color scheme may be developed that uses different colors to group various functions. Be certain that such techniques enhance the understanding of the reader; do not overdo it. Consistency across illustrations is important here, as changing color schemes from one drawing to the next may lead to confusion on the part of the reader.

Keep in mind that many readers may print the document on a black-and-white printer. For those situations, confirm that the illustration or photograph is still readable if printed monochrome.

When color is used, make an effort to keep the file sizes to a minimum for the sake of those who will download a PDF file containing the document. Screen shots can be saved in 16-color format to reduce file size. Some software publishing programs offer tools to optimize illustrations for screen viewing and printing.

3.3 Document Creation and Editing

Consistency throughout a document ensures a professional-looking finished product, and more importantly, minimizes the potential for confusion on the part of the reader. Abbreviations, acronyms, hyphenation, and capitalization should be uniform throughout the document. Headings, numbered and bulleted lists, tables, and reference lists should also be consistent.

Besides following the company style guide, writers should review the text for accuracy with subject matter experts (SMEs) during the writing project as well as before final publication of the documents.

3.3.1 Permissions

Authors and editors are strongly encouraged to prepare original figures and tables and avoid the use of borrowed material where possible. Submission of original materials eliminates the need to obtain permissions and facilitates revisions if needed. The following items can be copyrighted and their use requires permission:

- A table, diagram, or illustration (line drawing or photograph)
- A quote of 50 or more words from a periodical or journal
- A quote or series of shorter quotes totaling 400 words or more from a book

If you are the author of material copyrighted by another party, you must secure permission from that party to use the material in the new document. The important issue in determining whether permission is needed for an altered figure is the amount of alteration. The change must be substantial if you want to avoid the legal requirement to obtain permissions. What constitutes "substantial" change is a murky legal area. Changing straight lines to arrows, relabeling a figure with letters instead of numbers, or reordering columns in a table does not constitute substantial change and can distort the meaning of the original material. The best approach for avoiding permission issues is to use original materials wherever possible.

It is important to understand, however, that data cannot be copyrighted. Only the format in which it is published can be copyrighted. No permission is needed if data that appear in another text are converted to tabular form. If you are the first author to create a table comparing studies by four other scientists, you do not need permissions, but you should cite the studies as references.

Most printed materials of the U.S., Canadian, and British governments do not require permissions because they are in the public domain and not protected by copyright. However, many government-sponsored agencies copyright their materials, and use of such material requires permission. The best approach is to request permission unless you are certain that it is not required.

Permission requests are rarely denied, but they might be ignored. Some follow-up may be necessary. If permission cannot be obtained despite your best efforts, you can

- Delete the copyrighted material
- Find a substitute for the copyrighted material
- Substantially alter the material so permission is no longer required

A source line attributing material to a copyright holder who grants permission to use it should be included with the figure, photograph, or other material covered by the permission. The copyright holder may request special wording of the credit line.

3.3.2 Citing References

References, whether normative or informative, need to be properly documented. It is important to include as much information as possible in the reference listing in order to make sure that readers of the document can find the reference if they need to, sometimes many years (or even decades) after the original reference was published. Consistency in the reference listing is also important to readability.

Different publishers and organizations have their own guidelines for citing references. Whatever method is used, it should be documented and provided to the authors and editors of technical documents. For example, one general form for citing a reference is as follows:

[1] (author name): (name of document; if a published book use italics, if not use quotation marks), (name of publication and editor/author, if applicable), (publisher), (city), (state), (volume number, if applicable; e.g., vol. 4), (series number, if applicable; e.g., no. 1), (page number or range, if applicable; e.g., pp. 10–20), (date).

Citing a website reference typically requires a different treatment than citing a printed article or book. The availability of a web reference over time is another consideration as websites change and hyperlinks are broken. In any event, one general form for citing a web reference is as follows:

[1] (organization website name): (title of web page—if given—using quotation marks), (name of author— if given), (URL in the form <http//www.[URL and path as appropriate].extension>), (date).

References are typically numbered in the order in which they appear in the document. It is understood, however, that this practice may not always be practical, especially for documents that have been revised one or more times.

Within the body of the document, it is suggested that the reference citation use the document name/number followed the appropriate bracketed reference number. For example: "Constraints on the transport stream described in ATSC A/53 Part 2 [1] shall apply." It is further recommended that the revision letter or number of the reference not be included in the text, but rather in the reference listing. In this way, a new version of an existing document can be referenced simply by updating the references (and not every use in the text).

3.3.2.1 Examples

In some cases it may not be possible to capture all of the information suggested in this section for reference listings. Furthermore, certain types of documents will require special treatment. Still, the examples given in the previous section can serve as a starting point for developing complete and consistent references. Some examples are given below:

- For a published standard:

  [1] ATSC: "Digital Audio Compression Standard," Document A/52B, Advanced Television Systems Committee, Washington, D.C., June 14, 2005.

- For a chapter in a published book:

  [2] Donald G. Fink: "Video Colorimetry," in *Standard Handbook of Video and Television Engineering*, Jerry C. Whitaker and K. Blair Benson (Eds.), McGraw-Hill, New York, pp. 889–901, 2000.

- For a published book:

  [3] Peter Symes: *Video Compression*, McGraw-Hill, New York, 1999.

- For a magazine article:

  [4] Arthur Allison: "Making PSIP Work for You," *Broadcast Engineering*, Intertec Publishing, Overland Park, KS, vol. 35, no. 2, pp. 34–40, February 2000.

- For a web reference:

  [5] Advanced Television Systems Committee (ATSC): "Bylaws of the Advanced Television Systems Committee, Inc.," http://www.atsc.org/policy_documents/B-2-2008-04-09.pdf, April 9, 2008.

3.3.3 Terms

The issue of consistency across technical documents is important for a variety of reasons, not the least of which is to avoid confusion on the part of readers. The terms used to describe important concepts, variables, and constraints, therefore, are of critical importance. It is often useful to compile a Glossary of Terms that provides a master listing of all abbreviations and terms used in a technical document. To the extent possible, editors and authors of new documents should adopt naming conventions already in use for other documents.

Terms, abbreviations, and acronyms may be combined in a single list or separated out into dedicated lists. The approach chosen is often determined by the number of items that need to be listed in a given document. Definitions should be developed carefully, keeping in mind that a single term may be applied in different situations to describe different concepts. In a lengthy document, "term overload" can be a problem. This is where a particular term is used extensively, with more than one meaning. For example, the term "program" can be used to describe a project, a piece of content, a piece of computer software, etc. While the meaning of the term may be clear to the author of the document, it may be quite unclear to the reader.

3.3.4 Copyright, Trademark, and Legal Notices

Legal notices are required to protect company rights. Note that copyrights and trademarks do not necessarily have to be registered to be recognized. Contact the copyright office or a lawyer who specializes in copyrights and trademarks for specific advice. Distill the advice down to specific guidelines, and then apply the guidelines consistently across all documents and publications.

Trademarks may be acknowledged in text in one of two ways:

- Include the registered trademark symbol (^{TM}) and an asterisk in the text. Add the footnote, "*Registered trademark of company, city, state."

- Place the registration information in parentheses in the text along with the trademark symbol.

Capitalize subsequent mentions of a trademarked name. Some sample text is given below:

- Copyright: "This Guide and the related Software Product(s) are protected under a Copyright dated 2011 by Company Name. All rights are reserved."
- License agreement: "This Guide describes the installation and operation of Software Product, its environment and applications. It is made available only under the terms of a license agreement between the licensee and Company Name. No part of this Guide or the related Software Products(s) may be reproduced or transmitted in any form or by any means, electronic or mechanical, including photocopying and recording, for any purpose, without the express written permission of Company Name."
- Trademarks and copyrights of others: "The following products and/or registered trademarks of Company Name are referenced in this document: Product 1, Product 2, and Product 3."

3.3.5 Document Numbering

The numbering scheme used by a department or business group may vary; however, consistency is essential to permit efficient document access and version control. Any numbering system that works for an organization is acceptable. One possible approach is as follows:

g-xxxry-name.ext

where:
g   = group name abbreviation
xxx = the document number (three digits; e.g., 001, 002, etc.)
y   = the revision number (numeric), beginning with '0' as the first-release
name = brief descriptive text of the file using the hyphen character to separate words (no spaces)
ext   = the document extension.

For example: mfg-010r5-production-status-report.pdf.

Each time the document is edited or otherwise changed, the revision number should be incremented in order to facilitate version control.

It is often helpful to create and maintain a document log for all items—both internal and external. Likewise, a central repository for all documents, regardless of their type or intended use, is recommended.

Document archiving is often a task relegated to the back burner at many organizations. However, it is critically important for retracing previous steps, updating documents, product support, and related functions. It is best to have one person responsible for document archiving at an organization. This is practical in a small company or group but likely impractical for a large organization. However it is accomplished, a good document archive will pay for itself many times over.

3.3.6 Maintaining Corporate Image Consistency

There are a number of elements—both text and graphics—that convey the image of a given corporation or organization (Leitch, 2004). Keeping these elements consistent across a wide variety of publications is essential to maintaining the desired external image.

To promote the same "look and feel" to "brand" a company, product names, logos, taglines, unique spellings/punctuation, and other elements should always be followed—in the letterhead, documentation, marketing materials, website, and software product (display pages, error messages, and reports). Close attention should be paid to colors; standardize on colors, and insist on an exact match. Since some communication will be in black and white, include color and black and white specifications in the style guide.

Example guidelines are given below.

Next Generation 6000 Software—Technical Publications and Marketing Style Guidelines

Company name (always bold): **Next Generation 6000 Software, Inc.** or **NexGen6000 Software**. No spaces between "Nex," "Gen," and "6000." Always include "Software" (except on website, email addresses, and copyright fine print). First reference should include "Inc."

Product names (always bold). Abbreviations and special cases include the following:

- **NFO**—abbreviation of Next Focus. **NFO** should be all caps and bold.
- **Archive Recovery Environment** should be mixed case and bold.
- **ARE**—abbreviation of the Archive Recovery Environment. **ARE** should be all caps and bold.

User Guides: Always use bold and italics. Make sure the name in manual matches name on the title page of the manual. Example: *Next Focus Getting Started Guide*

Panel names: Always use bold and italics. Make sure the name in manual matches name on panel. Example: *Next Focus Primary Menu* web references: Use the following format, and black ink (rather than blue):

www.nexgen6000.com
Do not use http://www.nexgen6000.com; WWW.NEXGEN6000.
COM.

Title page:

- Title page will not be numbered and will not count in the pagination (e.g., page 0).
- Title page will include the company, product, and manual name.
- At the bottom-left of the title page, the following information is included in italics: release number, revision number, publication date.

Fonts:

- Title page: Arial font, 30 point, for company, product, and manual names
- Manual pages: Arial, 15 point, for page headings; Times New Roman, 12 point, for body; Times New Roman, 8 point, for far-left-side headings; Courier New, 7 point, for screen capture text.

Contact information: Include the corporate website in manuals and marketing collaterals. Include phone numbers and corporate address as appropriate. Include support@nexgen6000.com and info@nexgen6000.com as appropriate. Do not use specific employee contact information (name, e-mail address, phone number).

Guidelines such as these simplify the process of authoring and publishing documents. The best designs are usually simple ones, with a minimum number of fonts and graphic elements.

3.4 Corporate Identity

Branding is an essential element to an overall corporate strategy (Leitch, 2004). It is the outward view of a company's offering to the public and a key element in attracting and maintaining relationships with customers. Branding helps build cohesion among varied products and services. It ensures that customers recognize the value a company offers time and time again.

At the root of branding is the *core vision* of the company—the vision for which the brand is identified and linked with strategic company initiatives.

It defines

- Who we are as a company
- What we will be known for
- What our solutions offer
- How we communicate our brand

The core vision of the company is typically based on the *core values*, which can include some or all of the following:

- Individual accountability
- Flawless execution
- Integrity
- Innovative thinking
- Customer focus
- Collaboration

A corporate identity guidebook outlines all essential parts of a company's primary branding elements and describes how to optimize these elements to best promote brand recognition. Through proper use of the logo, colors, fonts, collateral, and internal elements, the brand becomes stronger and therefore, more compelling to current and future customers.

3.4.1 Logos

Clear guidelines need to be developed on the proper usage of a company's logos (Leitch, 2004). Consistent and correct usage of the logos and related branding elements are essential in communicating a cohesive, reliable solution. As old logos are updated or modified, these changes need to be reflected in all new supporting materials (letterhead, business cards, white papers, fact sheets, user guides, websites, e-mail communications, tradeshow booth artwork, mailings, etc.).

Deviation from the accepted version of a logo diminishes its value. Logo files should be provided for use by document authors and equipment panel designers (Figure 3.3).* They should not be recreated as needed.

To ensure the effectiveness of the logo, a minimum amount of space should be maintained around it. This space is also known as a "protected area." All elements such as taglines, images, other logos, headlines, names, or text must not enter this space. An example is given in Figure 3.4.

High contrast and visibility of the logo must be maintained over many different applications. Therefore, the logo may have several acceptable versions based on usage and background density. Some examples are given in Figure 3.5.

---

* Note: The company/brand names used in the illustrations and tables in this chapter are for example only. Any similarities to actual companies or products are coincidental.

FIGURE 3.3
A logo should be used consistently in documents and on products.

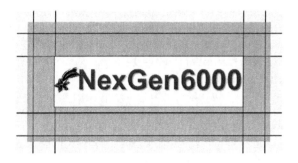

FIGURE 3.4
In this example, the protected area is defined as a specific height and width surrounding the logo. This area scales as the logo is scaled larger or smaller. The protected area is shown in gray.

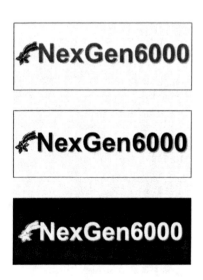

FIGURE 3.5
Logo variations: one-color versions of a logo are generally used for silk-screens, clothing, and special one-time uses; multicolor versions are used for all other applications.

It is also important to clearly document unacceptable uses of the logo. There may be specific placement or size considerations. For example, the logo of the parent company should never be smaller than the subsidiary company when used in the same application.

3.4.2 Colors

Colors are designed to enhance the company brand by invoking feeling and recognition in designs (Leitch, 2004). The colors provided in a style guide are usually given as guidelines for use; exact colors may vary depending on the limitations of the medium. These colors may be used for design elements, text (such as headlines), backgrounds, and accents. In addition, the specified colors are not the only design elements that may be used. Pictures, diagrams, and icons are all encouraged and are useful in conveying a message.

The recommended color palette may be split into multiple levels of use, depending on the planed application. A three-level system is common:

- *Primary colors* are those used most often in the company branding, typically reflecting the colors of the logo.
- *Secondary colors* are designed to be used in conjunction with the primary colors when another color is required for contrast.
- *Tertiary colors* are used only as highlights as part of a wider palette of colors that already includes primary and secondary colors. Tertiary colors can help focus designs when a specific market is being addressed.

A sample color palette is illustrated in Figure 3.6.

The color set is often specified according to their PMS (Pantone) color numbers for spot color printing. Visit Pantone's website for more information: www.pantone.com.

These colors may also be created by using 4-color process printing (CMYK) and on-screen (RGB). The values for CMYK and RGB should be provided for convenience.

3.4.3 Software Interfaces

Software interfaces are highly specialized for their audience. However, the same principles of consistency and colors may be applied. While form should follow function in the design of a user interface, keeping the look and feel of the interface similar to product literature is often beneficial to the user and to the company. Some sample interfaces are shown in Figure 3.7.

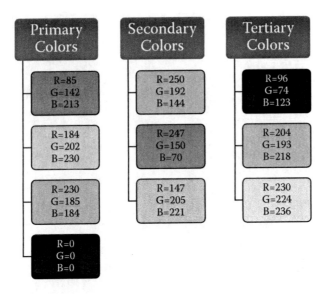

FIGURE 3.6
Example color palette, with three variations for each color group, plus black.

FIGURE 3.7
Sample web user interface that maintains the look of printed support materials.

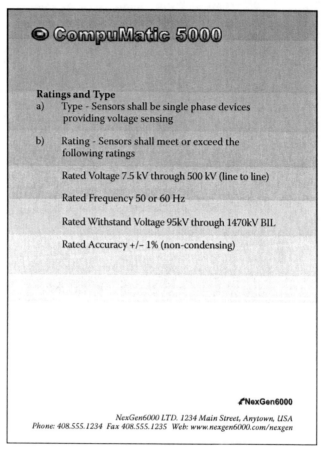

FIGURE 3.8
Sample advertising template.

3.4.4 Advertising Materials

The development of advertisements—print or otherwise—is a science in itself and is outside the scope of this book. However, many of the general guidelines provided in this chapter may also be applied to ads, in particular the importance of consistency.

An advertising template provides the starting point for compelling ads and maintains a graphic link with other supporting materials, including the look of the product itself. A template provides structure and serves to tie different products (or even different product lines) together graphically.

Figure 3.8 shows an example ad template. Note that the product logo is in the upper-left area and the company logo in the lower-right area. Telephone numbers and website reference are listed along the bottom.

3.5 Reference Books

It is recommended that readers utilize the following reference books for more information on basic writing standards:

Abrams, Eleanor S. et al. *The New York Public Library Writer's Guide to Style and Usage*, 1st ed. New York, NY: The Stonesong Press, Inc., 1994.

Microsoft Corporation. *The Microsoft Manual of Style for Technical Publications*, 2nd ed. Redmond, WA: Microsoft Press, 1998.

Sun Technical Publications. *"Read Me First!" A Style Guide for the Computer Industry*. Mountain View, CA: Sun Microsystems, Inc., 1996

The NYPL style book is a great reference for grammatical and editing decisions, while the MS style book is used for terminology, technical references, and usability advice. The SUN style book is helpful as a general reference. In situations where MS and/or SUN contradicts the NYPL style book on grammatical questions, you should follow the more scholastic NYPL.

References

Laplante, P.A., *Technical Writing: A Practical Guide for Engineers and Scientists*, CRC Press, Boca Raton, FL, 2011.

IEEE, "IEEE Recommended Practice for Software Requirements Specifications, Doc. 830-1993," Institute of Electrical and Electronics Engineers, New York, 1993.

Leitch, *Documentation Department Style Guide*, Edition A, Leitch Corporation, Toronto, Canada, 2004.

4

Meetings

4.1 Introduction

It is easy to poke fun at meetings. Sometimes they can be long, unproductive, and leave the participants wondering whether it was worth the time spent. However, meetings are critical to bringing together ideas from different perspectives. The key is to have *effective* meetings. While this may be easier said than done, this chapter contains some guidelines that will help to produce the desired result.

Meetings are important for an organization because products and processes are more complex today than ever before. There was a time not too long ago when a small group, or even a single person, could be assigned responsibility to develop, for example, a particular product. With the group located in the same office, meetings could be informal and done with little or no planning. Today, there are very few enterprises that can secure their future based on the vision of a handful of people who happen to be in the same geographic location. Meetings have become a necessity for most companies because the products those companies make are complex and often need to address different markets, perhaps in different parts of the world.

While it is satisfying for an individual or small group of workers to develop a new product or service on their own, the chances of success are limited because of the limited input such a process allows. It is fair to point out there are exceptions to this rule, some stunning; for example, two guys in a garage developing a small computer that changes the world. While these bursts of inspiration (and exceptional timing) still occur, the chances of success are quite limited.

Companies and organizations utilize meetings to bring together—in a structured manner—ideas from people who have different perspectives, backgrounds, goals, and objectives. Meetings also offer the opportunity for individuals to see the bigger picture they might otherwise miss because they are focused on their particular discipline, and not necessarily on the overall needs of the company.

Historically, when people think of meetings, they think of face-to-face meetings. That mindset has changed considerably over the years, thanks to improved communications tools. Organizations with widely spread members may use teleconferences for most meetings, coming together in person only at critical decision points. Among the driving factors for this approach are the improved efficiency of setting up and conducting the meetings, and cost savings. Travel can be extraordinarily expensive, particularly if it involves travel beyond geographical borders. The time required for travel is another matter entirely.

To be effective, meetings require a spirit of cooperation and the willingness to give in on points that may be important to one person or group but are not practical or appropriate for the greater good of the organization. Productive meetings begin with a positive attitude and the acknowledgement by all participants that no one person has all the answers. Cooperation and compromise are the keys to success.

4.2 Organization

An organization of any substantial size will invariably have a formal structure. Ideas can flow from the top down, or from the bottom up. In the ideal case, ideas will flow both ways. Likewise, ideas can come from inside the organization or from outside. Here again, under the best-case scenario, they come from all available sources. It is the responsibility of the organization to facilitate effective communications. One of the most common ways of accomplishing this is through meetings.

While a small group can meet informally with little or no advanced planning, such encounters lack a certain structure, and therefore the results may be lacking, or at least difficult to quantify. Formal meetings, on the other hand, are planned in advance with a clear agenda, set start and stop times, and specific goals in mind.

4.2.1 Responsibility of the Chairperson

The typical committee structure is illustrated in Figure 4.1. The chairperson is the administrative leader of the meeting. This person is charged with not only the responsibility to conduct the meeting, but also to manage the necessary work before and after the meeting. Planning for a meeting involves, at minimum, setting the agenda. Some agendas are simple; others are long and complex. More complex agendas require input from members of the committee, department heads, or others that will participate in the meeting.

The chairperson may choose to appoint a vice chair. In the chairperson's absence, the vice chair will temporarily assume the responsibilities of the chair. In addition, during the meeting there may be times where the chairperson will find the need to hand the gavel (and management of the meeting) to the vice-chair in order to make a personal statement on the subject being discussed.

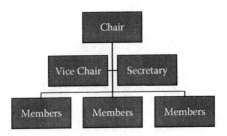

FIGURE 4.1
Typical committee structure.

The vice chair may also be assigned certain responsibilities by the chair, which may change from time to time. For example, the vice chair may assume responsibility for organizing the agenda for upcoming meetings, and for following up on action items after the meeting.

An effective chairperson will look for ways to move the discussion and the work forward, even in difficult situations where there exist entrenched positions. It is important for each viewpoint to be aired—each person should be given his or her opportunity to make a particular case. If strong opposing views are held by a small group of members, it may be productive to assign the task of coming up with an acceptable compromise to those individuals. If the members are willing to understand the viewpoints of others and compromise for the good of the organization, then progress can be made.

Sometimes the best approach to a problem is to defer a decision until later in the day or until the next meeting. A great deal of discussion takes place in the hallways outside meetings. Such private encounters can be quite productive in moving past sticking points.

In order to make sure major projects maintain forward momentum, the chairperson will want to assign action items to specific persons or groups. The action items need to be focused and should include a stated completion date, such as prior to the next meeting of the group.

4.2.1.1 Minutes of the Meeting

All standing committees should have a secretary to take minutes and track documents. The vice chair may also serve as the secretary. In the event that a group does not have a permanent secretary or if the secretary is not present, the chair may appoint any meeting attendee to act as secretary for that meeting. Minutes should be distributed in a timely fashion and approved by the attendees at the next meeting.

It is often difficult to find volunteers to serve as secretary, since the work can consume an inordinate amount of time. It may be necessary, therefore, to rotate the secretarial duties among committee members. The downside, however, is that the quality of the minutes of the meetings may be uneven.

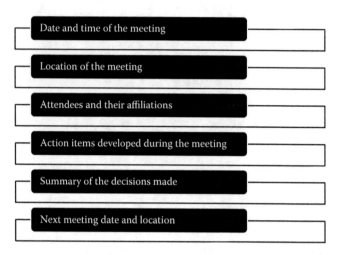

FIGURE 4.2
Essential elements of the minutes of a meeting.

Each organization may have its own approach to taking minutes. Still, a good starting point is illustrated in Figure 4.2. At a minimum, the minutes of a given meeting should include the following information:

- Date and times of meeting.
- Location of the meeting (if in-person).
- Attendees and their affiliation (or department, if appropriate). Make sure to include any guests present at the meeting.
- Action items developed at the meeting
- Summary of the decisions made at the meeting.
- Date and location of the next meeting.

Generally speaking, the best minutes are short minutes. It is easy to provide more detail than is really necessary in the minutes. The minutes should definitely not be a transcript of the meeting, but rather record the decisions made and actions taken. Some attendees may wish to have their particular position documented in the minutes. If useful for future decision-making, this approach may be acceptable. However, having the minutes document each person's pet position is rarely productive.

An approach that may be beneficial is for the chairperson to ask that anyone who wishes to document their position on a given point provide it in a written statement, which can be included with the minutes as an attachment. This keeps the minutes focused on actions taken, but at the same time provides a mechanism to record positions.

In an attempt to determine consensus on a given position, it may be useful for the chairperson to conduct a straw poll. Usually the results of such a straw poll are not recorded in the minutes, only whether the chairperson was able to determine consensus.

The draft minutes should be distributed to committee members in advance of the next meeting and then approved at the meeting, with any changes that might be required.

It may be useful for the secretary to tape-record the meeting in order to facilitate completing of the minutes. The recording, however, should be deleted when the minutes of the meeting are approved. There should be only one record of the meeting, and that is the approved minutes.

4.2.1.2 Scheduling Meetings

The chairperson is responsible for calling meetings at intervals necessary to handle the assigned work. There may be a fixed pattern to meetings (e.g., every Tuesday at 9:00 a.m.) or a flexible schedule, often dictated by key milestones set out in the Work Plan.

Meetings may be in-person, by telephone, or a combination of in-person and telephone. Using a web-sharing service where committee members can view a shared screen can be very effective in conveying ideas and reaching agreement on a particular strategy. Web-based desktop sharing is also quite useful for document editing by a group of experts.

Telephone and web-based meetings may also pose a scheduling conflict due to participants located in different time zones. The global economy means working with people in the United States, Europe, Asia, and South America. If San Francisco participants need to meet with engineers in Pune, India, this might mean an early morning call (India is 12.5 hours ahead of California).

One challenge often encountered for in-person meetings where some committee members participate by telephone is the distraction that can be caused by the telephone link and background noise problems. For in-person meetings, it may be advisable to have a standing policy that for in-person meetings, telephone connectivity is provided as a courtesy and if problems develop the meeting will continue. This avoids the possibility of a meeting being hijacked by troubleshooting phone bridge problems. One approach is given below:

> Participants are encouraged to attend meetings in person. At the discretion of the chairperson, dial-in teleconference capability may be provided. However, the quality of the system cannot be guaranteed and the use of the conferencing system will not unduly interfere with the meeting. The chair may discontinue use of the conferencing system if it becomes disruptive to the meeting.

Stating the rules up front tends to lower the level of aggravation when problems develop.

Sufficient advance notice of upcoming meetings is important to ensure good participation. Getting committee members to agree on a meeting schedule can be a challenge, however, given the complexities of travel and fixed or unpredictable events. Setting up a meeting schedule of once a week or twice a month—ideally at the same time of day—is often helpful in solving this challenge. Scheduling meetings well in advance can also be effective in keeping attendance high. It is not uncommon to schedule meetings out many months in advance, with the understanding that changes may need to be made based on unforeseen events.

When scheduling meetings, efforts should be made to avoid date and time conflicts with other company or related industry meetings. However, it is understood that this may not always be possible. The availability of meeting facilities must also be taken into account for in-person meetings.

For off-site meetings, the company will likely want to establish a policy on how costs incurred in attending the meeting are handled. Fortunately, improved telephone bridges, web-sharing services, and advanced teleconferencing systems have reduced the need for in-person meetings.

4.2.1.3 The Agenda

In preparation for an upcoming meeting, the chairperson should develop and distribute a draft agenda. The agenda is reviewed at the beginning of the meeting and revised if necessary by the group. As illustrated in Figure 4.3, the following items should be included on each agenda:

- Consideration of draft agenda
- Approval of minutes from the previous meeting
- Review of action items from the last meeting (action items can be marked *closed, carried forward,* or *overtaken by events*)
- Major discussion points
- Action items developed at the meeting
- Next meeting date, time, and location

If meetings are weekly or biweekly, having the minute approval process as part of next meeting works well. For less frequent meetings, it may be necessary to distribute the minutes and solicit comments via e-mail.

It is useful to develop a template for creation of agendas that is available to all committee chairs and vice chairs.

To facilitate productive meetings, documents and presentations to be discussed at a meeting should be distributed in advance. This gives participants the opportunity to review the material prior to the meeting and to formulate comments and suggest changes.

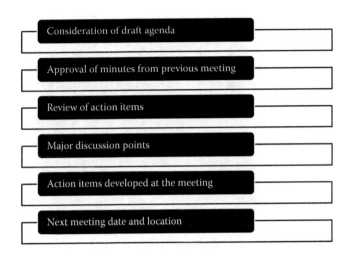

FIGURE 4.3
Basic items for a draft agenda.

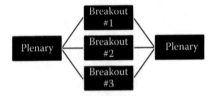

FIGURE 4.4
A meeting structure using breakout groups to address specific issues.

Various strategies are practical for organizing a meeting. One that works well for complex subjects is to organize break-out groups, where particular issues are discussed and solutions are identified (see Figure 4.4). Typically, the full committee will meet and take care of routine business. The committee then will break into smaller groups with specific objectives in mind. Later in the day, the full committee comes back together and reviews the recommendations of the break-out groups. The full committee then forms a final overall plan or strategy based on this input.

A policy should be established for each committee regarding attendance by individuals who are not members of the committee. Typically, meetings are open to anyone with an interest in the work. However, it is understood that some projects require confidentiality. It is always best to identify such issues before the meeting starts, since asking someone to leave a meeting is awkward for both parties.

In some cases, meeting participants may be asked to sign a nondisclosure agreement. This can raise difficult questions that are best addressed well before the meeting begins.

Presentation material should also be prepared with the meeting in mind. For example, if confidential or proprietary material is contained in a presentation, then it should not be shown in an open meeting. Many companies have presentation (PowerPoint) templates that contain "confidential" or "proprietary" text in the footer. If a presentation marked "confidential" is shown to a public group, legal issues may arise.

When the decision of a committee needs to be communicated to others in the organization, it should be done by the chairperson or someone designated by the chairperson to report on behalf of the committee. Having multiple versions of what was decided by the group is a recipe for problems.

4.2.1.4 Quorum

Rules regarding the establishment of a quorum for a meeting range from very simple to complex. In the simplest case, there is no formal determination of quorum. The meeting proceeds with whomever shows up. While this approach certainly has the benefit of simplicity, it can also lead to decisions being made that do not accurately reflect the views of all stakeholders. Furthermore, for an effective meeting a certain critical mass of participants is needed; otherwise, the committee tends to just spin its wheels.

At the other end of the spectrum, quorum may be formally recorded at each meeting based on various criteria, such as the number of voting-eligible members present. While quorum is typically 50 percent plus one, some other benchmark may be set by the organization. Whatever method is used, the rules regarding quorum should be documented.

There may be meetings where decisions must be made by a "consensus body" that is representative of the membership of the organization. In such cases, the organization may establish voting eligibility rules, which might require attendance at some minimum number of past meetings (e.g., two of the last three meetings of the group). The goal here is to make sure persons voting on critical matters have been involved in the work of the committee and so presumably fully understand the decisions being made.

4.2.2 Challenges of Working Remotely

Our society is accustomed to staying connected with others—we talk on our cell phones when driving, we use our smart phones or laptops when waiting for a plane, and we use our cell phones to send a photo of our son playing basketball.

In the workplace, we may interact with co-workers, contract employees, and customers around the world. With increasing airfares and decreasing travel budgets, we may be required to conduct business remotely with colleagues residing in the same state or province, or anywhere around the world.

More and more business communications take place via e-mail because it is cheaper and more time-effective than phone calls. File sharing and collaborating is a fairly new concept, but more prevalent today. Teleconferences (conference calls) and web meetings (via the Internet) may replace the traditional face-to-face meetings. Web meetings are becoming increasingly popular not only for company meetings but also for meetings with customers and prospective customers.

These electronic communication methods can be very effective, but they are not perfect (or should we say the people that use them are not perfect). We list a few suggestions here to help embrace the technologies and overcome the obstacles.

4.2.2.1 Time Zones, Daylight Savings, Holidays

Working with people in different time zones can be a challenge. Meetings planned for 9:00 a.m. California time would be at noon in New York. So you will want to be sensitive to the various remote locations that will be joining your meeting. Scheduling can be especially challenging if you are on the West Coast and needing to connect with people in Europe and Asia-Pacific (you may need to split into two separate meetings). Two websites that can help are the World Clock (see Figure 4.5) and the World Meeting Planner (see Figure 4.6). Both of these are available as iPhone apps.

The World Clock website can help if you are planning a meeting or phone call with one other location (simply scan the listed cities for the one you're interested in calling). The World Meeting Planner website is especially helpful when planning meetings with locations in more than one time zone. For example, enter Los Angeles, New York, and Rome on the webpage. The results show a list of potential meetings—times highlighted in green would be within normal business hours (see Figure 4.7). Sometimes meetings must be held outside of the standard work-day at both ends. The 12.5-hour time difference between California and India makes it impossible for all attendees to meet within normal business hours.

Daylight Savings Time (DST) was introduced to make better use of the daylight hours during the summer months. About 70 countries observe some form of DST. Countries south of the Equator will not observe DST during the same months as the countries to the north (summer months are opposites). Additionally, the countries along the Equator and within the Tropics do not observe DST because their daylight hours do not vary much throughout the year.

Since DST differs from country to country (and state to state), it is best to verify when scheduling phone calls and meetings. DST is shown on the World Clock and World Meeting Planner websites.

Holidays are another factor to be considered when scheduling a meeting. Within the United States, most holidays are observed on a Monday or a Friday (individual states may have their own holidays which may or may not fall

FIGURE 4.5
World Clock (time zones) website (http://www.timeanddate.com/worldclock/).

on a Monday or Friday). Countries can have a wide variety of national and religious holidays. If you have regularly scheduled meetings (on a weekly or biweekly basis), these holidays need to be taken into consideration. The Internet provides a good resource for this information. For example if you're working with a team in India, Google "India holidays 2012". You will find several "hits" listing the bank and public holidays. Verify with your team the holidays they observe (the team may be impressed and appreciate your interest in their country).

4.2.2.2 Language Barriers

Communicating with people who speak the same language is often challenging, but when you are trying to communicate complex business or mechanical designs to someone from another country, it can be even more

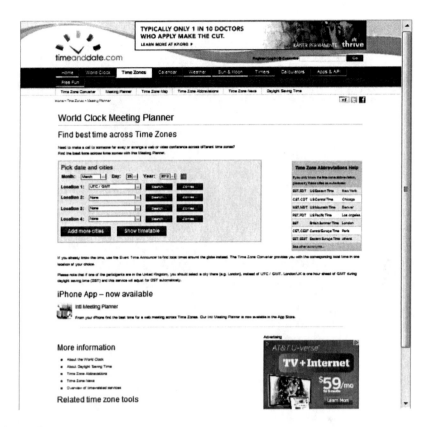

FIGURE 4.6
World Meeting Planner website (http://www.timeanddate.com/worldclock/meeting.html).

challenging. We take it for granted that people will understand us, but we need to choose our words carefully. When writing documentation that will be translated, use simple terms. The same policy should be followed in your conversations and meetings.

Do not use clichés or slang phrases. "The customer shot himself in the foot" may cause a non-native English-speaking co-worker to become upset. Of course, it means that the customer had made a mistake that caused a big problem. Although these can be a simple way to communicate an idea, a better approach would be painting this image through effective writing. The easiest modification would be to replace the slang phrase with a word or phrase that has the same meaning. A few examples are found in Table 4.1.

Patience and courtesy are important in dealing with language issues during a meeting. Remember that participants in other countries are trying hard to communicate with you even though English (or some other language) may not be their native tongue. Resist the urge to become frustrated when you cannot understand the speaker. It is often helpful to repeat back what you think you heard; for example, "I understand that you are proposing we do X

FIGURE 4.7
World Meeting Planner results (LA, NY, Rome).

TABLE 4.1

Clichés and Possible Replacements

Cliché	Possible replacement
Bite the bullet	Sacrifice
Don't mince words	Be precise
Drop in the bucket	Tiny
Feather in your cap	Accolade
Happier than a clam at high tide	Satisfied
In this day and age	Today
Ins and outs	Details
Nitty-gritty	Details
No mean feat	Difficult
Only time will tell	Eventually
The world is your oyster	An opportunity arises
Whets the appetite	Invites

Source: Laplante, P.A., *Technical Writing: A Practical Guide for Engineers and Scientists*, CRC Press, Boca Raton, FL, 2011.

and then do Y. Is that correct?" This helps move the meeting along and demonstrates that you are working with the speaker to make certain everyone understands his or her position.

Imagine if the roles were reversed and you were the one speaking in a foreign language. You would want others in the meeting to be patient and courteous.

4.2.2.3 Cultural Differences

Similar to language differences, cultural differences can cause problems in communications. What may seem obvious to you may not be the standard practice in another country. If you can try to understand some of the cultural differences and carefully document your expectations, you will minimize the potential for problems.

Culture can also dictate the normal business hours for the country. For example, in Spain workers start in the morning—working until mid-day. They return home for a couple of hours, having lunch and relaxing with their families. It's rude to call during this time (so do not schedule a remote meeting). Afterward, they return to work. Typical dinner time is after 9:00 p.m.[*]

4.2.2.4 Voice Communications

Today, there is a myriad of choices for long-distance voice communications. Although landlines are still prevalent for businesses, Voice over IP (VoIP) is gaining popularity. There are also companies that specialize in conference calling plans. Some services are better than others. When considering a new phone system or calling plan, contact persons in other departments or organizations that have experience in the candidate systems. For teleconference meetings, voice quality and reliability are very important.

4.2.2.5 E-mail Communications

Electronic mail (e-mail) messages can be an effective way to communicate with colleagues (Laplante, 2011). These messages can convey the same type of informality as a face-to-face conversation or the formality of a legal document. Since e-mails can be saved for an extended period of time, it is best to carefully word them (do not reply to an e-mail when you are angry). In-person or phone conversations may warrant a follow-up e-mail message (confirming understanding or actions to be taken). Remember, e-mails cannot communicate your tone or body language.

In the carpentry trade there is an old saying, "measure twice, cut once." This rule should also apply to e-mail messages—read twice before sending. Remember that an e-mail can live forever on a server somewhere, and may find its way to persons far removed from the intended recipient. As such, it

---

[*] Reference: http://wiki.answers.com/Q/Difference_between_US_culture_and_Spain_culture.

is important to make sure that confidential information is not disclosed or damaging statements are not made about a person, project, or organization. Such messages may come back to haunt you.

The following are suggested guidelines for professional e-mails:

- Keep messages short and concise (avoid flowery language and needless complexity).
- Use relevant keywords in the subject line (helpful for searching and organizing e-mails).
- Include a question, call to action, or brief explanation of an attached document.
- Specify deadlines for any calls to action.
- Include a "signature line" with your name, title, company, and the best ways to contact you (e-mail address and phone number).
- Avoid the "receipt confirmation" option (may be perceived as annoying by the recipient).
- Avoid setting the "urgent" message flag (urgent e-mails should probably be phone calls).

For group work involving a large number of participants, one or more e-mail reflectors may be set up. Such tools are helpful in keeping everyone updated on the status of a project. When posting (or responding) to a reflector message, it is important to choose your words carefully. One problem commonly encountered with a reflector is responding to the group versus responding to an individual. You don't have to look far to find instances where someone thought they were responding to an individual, but mistakenly copied the entire group. This can result in an awkward situation.

Sometimes the best response to an e-mail is no response at all. It is easy to get a tit-for-tat e-mail fight going, where the persons involved feel the need to respond to every criticism or negative comment posted. Following up on messages is a good thing, but sometimes the better approach is to just let an issue die of neglect. Simply "getting the last word" in an e-mail fight is of little value in business today.

For sensitive material or personnel matters, it may be best to avoid e-mail altogether, instead using in-person or telephone conversations. Remember that e-mail is not necessarily private or privileged communications.

4.2.2.6 Virtual or Remote Meetings

Virtual or remote meetings can be facilitated via a web conferencing product. Web conferencing can connect you "live" with people almost anywhere on the planet. The connection can include slide presentations (PowerPoint), desktop sharing, and live or streaming video. There are also audio options (either via computer speakers/microphone or phone connection). Web conferencing providers include Skype, WebEx, GoToMeeting, Acrobat Connect, and others.

Free services such as Skype provide a conduit to hold lengthy video or teleconference calls halfway around the world at no cost to the participants. Voice and image quality can be impacted considerably by the data rate of the subscriber connection, and as such the connection may not always be consistent. As service providers continue to improve their systems, these issues will recede. In addition, such services are increasingly offering the capability to share data files as well as typical audio/video conference call functions.

WebEx and similar services combine desktop sharing (including PowerPoint slideshows) through a web browser interface. With phone conferencing and video, it is like everyone being in the same room. Even the desktop control can be transferred to a remote participant. These services continue to add features that are useful to meeting participants. As the complexity increases, however, it is important that the meeting chair (or another designated person) understands how to run the system. It is nonproductive to spend the first ten minutes of a meeting trying to figure out how to use the tools.

4.2.2.7 File Sharing/Collaborating

There are several software packages that can help you to manage documentation and collaborate remotely. Review of these packages is beyond the scope of this book, but here is a brief list of some of the more popular options:

- *Microsoft SharePoint* product allows users to set up websites to share information with others, manage documents, and publish reports. (See http://www.sharepoint.microsoft.com.)
- *Microsoft Project* product provides users with tools for unified portfolio and project management. (See http://www.microsoft.com/project/.)
- *Adobe Acrobat.com* can be used to host online meetings, store files online, and collaborate on documents. (See http://www.acrobat.com.)
- *Google Docs* provides users with a tool set for basic projects. (See http://www.docs.google.com.)

There are other vendors in this space, giving end users a number of options from which to choose. As with web-based meeting tools, it is important to understand these products in order to use them effectively. Generally speaking, the organization should select and standardize on a set of common tools across the company or group. This will eliminate the necessity of having members learn how to use new tools as the first step in a new project.

4.2.2.8 Less Formal Collaboration

Large projects may benefit from less formal communications (Laplante, 2011). Implementation of a newsletter, website, or blog can help foster a sense of community. Newsletters are informal publications that have a fast turn-around and can expose an idea for rapid consideration and discussion. Similarly, websites and blogs can be very timely in their coverage and solicitation of feedback. These communications could help reduce the number of meetings required for the project, and might help the meetings to run smoother. They may also serve to build interest and enthusiasm for a particular project or organization-wide effort. People like working on projects that have visibility; a newsletter and blog are effective ways of increasing the visibility of a project.

As practical matter, it can be difficult to produce a quality newsletter over a long period of time. It is important, therefore, to involve a number of persons within the organization in drafting articles, providing photographs, and otherwise contributing to the publication. By spreading the task out among a number of persons, the overall job becomes considerably easier, and the end product considerably better.

4.2.3 Developing a Scope of Work

Each standing committee should develop and agree to a formal Scope of Work. The best scope statements are clear, direct, and concise. The purpose of the scope is to guide the work of the committee and to avoid the tendency of committee members to wander into areas that are either covered by another committee, or in the view of the group or organization are not productive. In the event of a structured committee architecture, where subgroups report up to a parent group, the parent committee should approve the scope statements of each of its subgroups.

Because the needs of a company or organization change over time, the Scope of Work should be reviewed periodically, typically on an annual basis.

4.2.4 Developing a Work Plan

A Work Plan describes key objectives and the timeline for completing them. It is understood that a Work Plan may change over time as the work of the group progresses. A Work Plan need not be detailed or complex, but should—at a minimum—identify important milestones and attach achievable dates to them.

The Work Plan must reflect the realities of the resources available to the committee and to its members. It is of little value to set target dates that in all likelihood cannot be met. By the same token, committee work tends to be driven by schedules. Without target dates for completion, progress can be difficult to achieve.

4.3 Decision Making

In an ideal world, the members of the committee will all work toward the stated goals of the organization and leave their individual agendas at the door. This ideal situation can be hard to find, however, and so a formal structure is usually needed for determining how decisions are made in a committee.

Most decisions in most committees are probably made by the consensus of the group. Consensus can be defined differently by different organizations, but generally speaking it means a clear majority of the members present. Consensus does not mean unanimity. Differences of opinion may still exist; however, the clear majority of the group is in favor of (or opposed to) some stated course of action.

Decisions can also, of course, be made by a formal vote of the members of the committee. This approach requires advance planning to determine the basis rules, which include (but are not limited to) the following:

- *Which committee members have voting privileges?* The options include (1) everyone attending the meeting has a vote, (2) one vote per department (or company in the case of a member organization consisting of company representatives), or (3) some other criteria.

- *Are there voting eligibility requirements?* In order to have voting status, some organizations specify that the member must have attended some minimum number of previous meetings; for example, two of the last three scheduled meetings. This approach requires a certain amount of administrative overhead and speaks to the need for accurate meeting attendance lists.

- *Does the chairperson have a vote?* The options include (1) yes, all the time, (2) votes only in the event of a tie, (3) does not vote, (4) makes the final decision without any formal vote of the members, or (5) some other arrangement.

- *What is the passing percentage?* The two common options are (1) 50 percent plus one or (2) 2/3rds majority. Generally speaking, the voting percentage is determined by the members present at the meeting with authority to vote, not counting abstentions.

There are other considerations as well, depending on the makeup of the committee and the overall organization. Regardless of the structure, it is very important to develop a clear statement on the process that will be followed with regard to voting, and to make that information available to all members. Trying to sort out voting issues during a meeting when a vote is being taken is always a bad practice.

One option for the chairperson is to operate by consensus whenever possible, but resort to a "two-thirds rule" if a clear consensus cannot be reached.

In such a case, the decision of the group is determined by a vote of at least two-thirds of the eligible members.

Committee members objecting to some action by the group are usually afforded the opportunity to prepare a minority report or statement on a given decision. This option is rarely exercised, except in the case of very contentious issues.

One obstacle to making an important decision is having the right people in the room when the decision is made. This can be particularly troublesome if a key stakeholder is absent. A useful alternative to simply delaying the decision is for the group to arrive at a Tentative Decision, with the final decision to come at the next scheduled meeting. All members are notified of the Tentative Decision immediately after the meeting, giving them clear notice of the intended action at the next meeting. Members then have the opportunity to make comments or suggest changes if they wish.

From time to time it is likely that difficult issues will come up that cannot be resolved during the meeting. One method of moving forward that can be productive is to form a small ad hoc group consisting of members with differing views on the subject and task them with coming to some agreement among themselves, and then bring back a recommendation to the full group. It is usually best to narrow the scope of the ad hoc group to focus on specific sticking points rather than broad concepts.

Sometimes the obstacle to approval of a document or proposal by the committee has more to do with the visibility of the proposal than the proposal itself. For example, a proposal put before the committee with essentially no advance notice may face a tough time being approved. Options include deferring a decision until the next meeting or approving the document with the provision that if any member objects within a specified review period (e.g., two weeks), the document will need to come back to the full committee.

4.4 It's All about Leadership

The range of issues that the leadership of a committee may need to deal with is broad and largely unpredictable. However, armed with clear procedures (and a copy of *Robert's Rules of Order*) progress can still be made on difficult issues. *Robert's Rules of Order* is typically the default "governing document" for most standing committees in the event that procedures specific to a particular committee have not otherwise been defined. With planning and good common sense, the need for having a copy of "the Rules" or the organization's procedures document on the table during a meeting will rarely present itself.

Whatever process is used for meetings, the rules should be documented and made available to all participants well in advance of the meeting. It is easy for a committee to get distracted with procedural issues that have little or no benefit to the task at hand.

As a practical matter, a good meeting is one that produces useful results. More generally speaking, however, the best meetings typically have the following attributes:

- Agenda distributed well in advance of the meeting.
- Appropriate documents distributed well in advance of the meeting.
- All key stakeholders show up.
- The meeting begins on time.
- Presentations and briefings are clear and to the point.
- Where specific actions are required, they are clearly outlined to the members.
- Decisions are made as needed.
- Action items are recorded and assigned to specific members with agreed-upon due dates.
- Concise minutes of the meeting are recorded.
- The meeting ends on time.
- Follow-up actions after the meeting are completed.

When done right, meetings can serve as an essential element of organizational development and can even be an enjoyable experience for those involved.

References

Laplante, P.A., *Technical Writing: A Practical Guide for Engineers and Scientists*, CRC Press, Boca Raton, FL, 2011.

5

Systems Engineering*

5.1 Introduction

Modern *systems engineering* emerged during World War II as weapons evolved into weapon systems—due to the degree of complexity in design, development, and deployment. The complexities of the space program in the 1960s made a systems engineering approach to design and problem-solving even more critical. Indeed, the Department of Defense and NASA are two of the staunchest practitioners of systems engineering. With the growth of digital systems, the need for systems engineering gained even more attention as new technologies were introduced at a rapid rate.

Today, most large engineering organizations utilize a systems engineering process. Much has been published about system engineering practices in the form of manuals, standards, specifications, and instructions. In 1969, MIL-STD-499 was published to help the government and contractors interface in support of defense acquisition programs. In 1974, this standard was updated to MIL-STD-499A, which specifies the application of system engineering principles to military development programs. The tools and techniques of this process have continued to evolve since, saving time and cutting costs.

5.2 Systems Theory

Systems theory is readily applicable to engineering of control, information processing, and computing systems. These systems are made up of component elements that are interconnected and programmed to function together. Many systems theory principles are routinely applied in the aerospace, computer, telecommunications, transportation, and manufacturing industries.

---

* This chapter is based on DeSantis, G., "Systems Engineering," in *The Electronics Handbook*, ed. Jerry C. Whitaker, CRC Press, Boca Raton, FL, 2005.

FIGURE 5.1
The major elements of systems theory.

For the purpose of this discussion, a *system* is defined as a set of related elements that function together as a single entity.

Systems theory consists of a body of concepts and methods that guide the description, analysis, and design of complex entities (see Figure 5.1).

Decomposition is an essential tool of systems theory. The systems approach attempts to apply an organized methodology to completing large complex projects by breaking them down into simpler and more manageable component elements. These elements are treated separately, analyzed separately, and designed separately. In the end, all of the components are recombined to build the whole.

Holism is an element of systems theory in that the end product is greater than the sum of its component elements. In systems theory, the modeling and analytical methods enable all essential effects and interactions within a system and those between a system and its surroundings to be taken into account. Errors resulting from the idealization and approximation involved in treating parts of a system in isolation, or reducing consideration to a single aspect, are thus minimized.

Another holistic aspect of systems theory describes *emergent properties*. Properties that result from the interaction of system components, that are not those of the components themselves, are referred to as emergent properties.

Though dealing with concrete systems, *abstraction* is an important feature of systems models. Components are described in terms of their function rather than in terms of their form. Graphical models such as block diagrams, flow diagrams, timing diagrams, and the like are commonly used.

Mathematical models may also be employed. Systems theory shows that, when modeled in abstract formal language, apparently diverse kinds of systems show significant and useful *isomorphism* of structure and function. Similar interconnection structures occur in different types of systems. Equations that describe the behavior of electrical, thermal, fluid, and mechanical systems are essentially identical in form.

Isomorphism of structure and function implies isomorphism of the behavior of a system. Different types of systems exhibit similar dynamic behavior, such as response to stimulation.

The concept of hard and soft systems appears in systems theory. In hard systems, the components and their interactions can be described by mathematical models. Soft systems cannot be described as easily. They are mostly human activity systems, which imply unpredictable behavior and nonuniformity. They introduce difficulties and uncertainties of conceptualization, description, and measurement. The kinds of system concepts and methodology described previously cannot be applied.

5.2.1 Systems Engineering Process

Systems engineering depends on the use of a process methodology based on systems theory. To deal with the complexity of large projects, systems theory breaks down the process into logical steps.

Even though underlying requirements differ from program to program, there is a consistent, logical process that can best be used to accomplish system design tasks. The basic product development process is illustrated in Figure 5.2. Systems engineering, at the beginning of this process, describes the product to be designed. It includes four major activities:

- Functional analysis
- Synthesis
- Evaluation and decision
- Description of system elements

This process is iterative, as shown in Figure 5.3. With each successive pass, the product element description becomes more detailed. At each stage in the process, a decision is made whether to accept, make changes, or return to an earlier stage of the process and produce new documentation. The result of this activity is documentation that fully describes all system elements and that can be used to develop and produce the elements of the system. The systems engineering process does not produce the actual system itself.

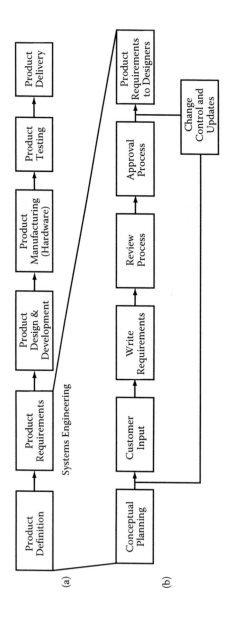

FIGURE 5.2
Systems engineering: (a) product development process; (b) requirements documentation process.

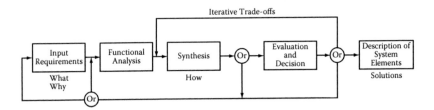

FIGURE 5.3

The systems engineering design process. (Adapted from Hoban, F.T., and W. M. Lawbaugh, *Readings in Systems Engineering*, National Aeronautics and Space Administration, Washington, DC, p. 12, 1993.)

5.2.2 Functional Analysis

A systematic approach to systems engineering will include elements of systems theory (see Figure 5.4). To design a product, hardware and software engineers need to develop a vision of the product: the product requirements. These requirements are usually based on customer needs researched by a marketing department. An organized process to identify and validate customer needs will help minimize false starts.

System objectives are first defined. This may take the form of a *mission statement*, which outlines the objectives, the constraints, the mission environment, and the means of measuring mission effectiveness. The purpose of the system is defined, and analysis is carried out to identify the requirements and what essential functions the system must perform and why.

The functional flow block diagram is a basic tool used to identify functional needs. It shows logical sequences and relationships of operational and support functions at the system level. Other functions, such as maintenance, testing, logistics support, and productivity, may also be required in the functional analysis. The functional requirements will be used during the synthesis phase to show the allocation of the functional performance requirements to individual system elements or groups of elements. Following evaluation and decision, the functional requirements provide the functionally oriented data needed in the description of the system elements.

Analysis of time-critical functions is also a part of this functional analysis process when functions have to take place sequentially, concurrently, or on a particular schedule. Time-line documents are used to support the development of requirements for the operation, testing, and maintenance functions.

5.2.2.1 Synthesis

Synthesis is the process by which concepts are developed to accomplish the functional requirements of a system. Performance requirements and constraints, as defined by the functional analysis, are applied to each individual element of the system, and a design approach is proposed for meeting the

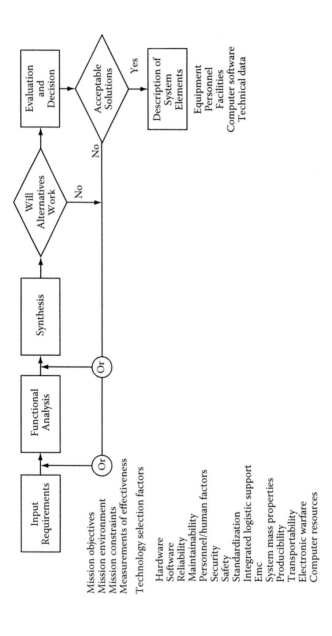

FIGURE 5.4

The systems engineering decision process. (Adapted from Hoban, F.T., and W. M. Lawbaugh, *Readings in Systems Engineering*, National Aeronautics and Space Administration, Washington, DC, p. 12, 1993.)

requirements. Conceptual schematic arrangements of system elements are developed to meet system requirements. These documents can be used to develop a description of the system elements and can be used during the acquisition phase.

5.2.2.2 Modeling

The concept of modeling is the starting point of synthesis. Since we must be able to weigh the effects of different design decisions in order to make choices between alternative concepts, modeling requires the determination of those quantitative features that describe the operation of the system. We would, of course, like a very detailed model with as much detail as possible describing the system. Reality and time constraints, however, dictate that the simplest possible model be selected to improve the chances of design success. The model itself is always a compromise. The model is restricted to those aspects that are important in the evaluation of system operation. A model might start off as a simple block diagram with more detail being added as the need becomes apparent.

5.2.2.3 Dynamics

Most system problems are dynamic in nature. Signals change over time and the components determine the dynamic response of the system. The system behavior depends on the signals at a given instant, as well as on the rates of change of the signals and their past values. The term *signals* can be replaced by substituting human factors such as the number of users on a computer network, for example.

5.2.2.4 Optimization

The last concept of synthesis is optimization. Every design project involves making a series of compromises and choices based on relative weighting of the merit of important aspects. The best candidate among several alternatives is selected. Decisions are often subjective when it comes to deciding the importance of various features.

5.2.3 Evaluation and Decision

Program costs are determined by the trade-offs between operational requirements and engineering design. Throughout the design and development phase, decisions must be made based on evaluation of alternatives and their effect on cost. One approach attempts to correlate the characteristics of alternative solutions to the requirements and constraints that make up the selection criteria for a particular element. The rationale for alternative choices in the decision process is documented for review. Mathematical models and computer simulations may be employed to aid in the evaluation and decision-making process.

5.2.3.1 Trade Studies

A structured approach is used in the trade study process to guide the selection of alternative configurations and ensure that a logical and unbiased choice is made. Throughout development, trade studies are carried out to determine the best configuration that will meet the requirements of the program. In the concept exploration and demonstration phases, trade studies help define the system configuration. Trade studies are used as a detailed design analysis tool for individual system elements in the full-scale development phase. During production, trade studies are used to select alternatives when it is determined that changes need to be made. Figure 5.5 illustrates the relationship of the various types of elements that may be employed in a trade study.

Figure 5.6 is a flow diagram of the trade study process. To provide a basis for the selection criteria, the objectives of the trade study must first be defined. Functional flow diagrams and system block diagrams are used to identify trade study areas that can satisfy certain requirements. Alternative approaches to achieving the defined objectives can then be established.

Complex approaches can be broken down into several simpler areas, and a decision tree constructed to show the relationship and dependences at each level of the selection process. This *trade tree*, as it is called, is illustrated in Figure 5.7. Several trade study areas may be identified as possible candidates for accomplishing a given function. A trade tree is constructed to show relationships and the path through selected candidate trade areas at each level to arrive at a solution.

Several alternatives may be candidates for solutions in a given area. The selected candidates are then submitted to a systematic evaluation process intended to weed out unacceptable candidates. Criteria are determined that are intended to reflect the desirable characteristics. Undesirable characteristics may also be included to aid in the evaluation process. Weights are assigned to each criterion to reflect its value or impact on the selection process. This process is subjective. It should also take into account cost, schedule, and hardware availability restraints that may limit the selection.

The criteria data on the candidates are then collected and tabulated on a decision analysis worksheet (Figure 5.8). The attributes and limitations are listed in the first column, and the data for each candidate listed in adjacent columns to the right. The performance data may be available from vendor specification sheets or may require laboratory testing and analysis. Each attribute is given a relative score from 1 to 10 based on its comparative performance relative to the other candidates. Utility function graphs (Figure 5.9) can be used to assign logical scores for each attribute. The utility curve represents the advantage rating for a particular value of an attribute. A graph

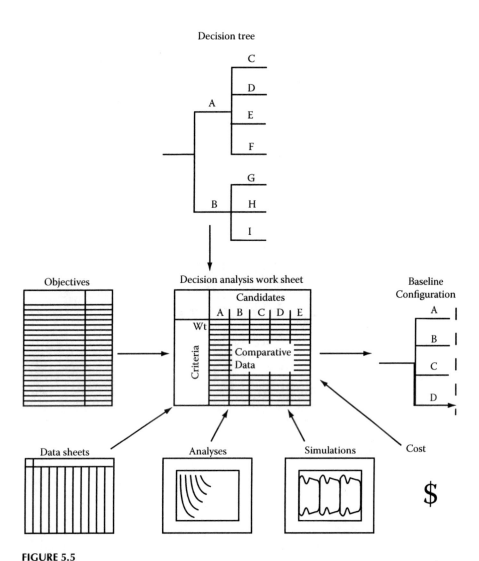

FIGURE 5.5
Trade studies using a systematic approach to decision making. (Adapted from *Defense Systems Management*, Systems Engineering Management Guide, Contract No. MDA 903-82-C-0339, Defense Systems, Management College, Fort Belvoir, VA, 1983.)

is made of ratings on the y axis versus attribute value on the x axis. Specific scores can then be applied, which correspond to particular performance values. The shape of the curve may take into account requirements, limitations, and any other factor that will influence its value regarding the particular criteria being evaluated. The limits to which the curves should be extended should run from the minimum value, below which no further benefit will accrue, to the maximum value, above which no further benefit will accrue.

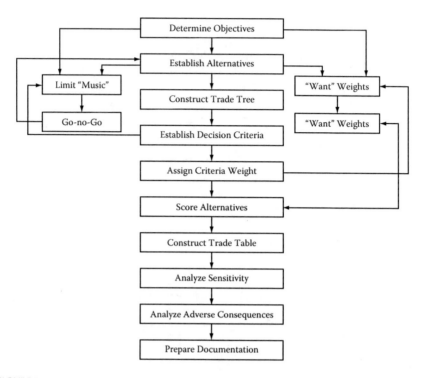

FIGURE 5.6
Trade study process flowchart. (Adapted from *Defense Systems Management*, Systems Engineering Management Guide, Contract No. MDA 903-82-C-0339, Defense Systems, Management College, Fort Belvoir, VA, 1983.)

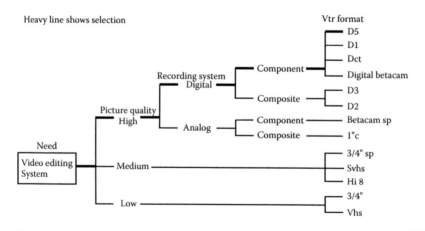

FIGURE 5.7
An example trade tree.

Alternatives:		Candidate 1			Candidate 2			Candidate 3		
Wanted	Wt		Sc	Wt Sc		Sc	Wt Sc		Sc	Wt Sc
Video bandwidth mhz	10	5.6	10	(100)	6.0	10	(100)	5.0	9	90
Signal-to-noise ratio, db	10	60	8	80	54	6	60	62	10	(100)
10-bit quantizing	10	Yes	1	10	Yes	1	10	Yes	1	10
Max program length, h	10	2	2	20	3	3	(30)	1.5	1.5	15
Read before write capable	5	Yes	1	(5)	Yes	1	(5)	No	0	0
Audio pitch correction avail	5	Yes	1	(5)	No	0	0	Yes	1	(5)
Capable of 16:9 aspect ratio	10	No	0	0	Yes	1	(10)	Yes	1	(10)
Employs compression	-5	Yes	1	-5	No	0	(0)	Yes	1	-5
Sdif (serial digital interface) built in	10	Yes	1	(10)	Yes	1	(10)	Yes	1	(10)
Current installed base	8	Medium	2	(16)	Low	1	8	Low	1	8
Total weighted score:				241			234			(243)

FIGURE 5.8

Decision analysis worksheet example. (Adapted from *Defense Systems Management*, Systems Engineering Management Guide, Contract No. MDA 903-82-C-0339, Defense Systems, Management College, Fort Belvoir, VA, 1983.)

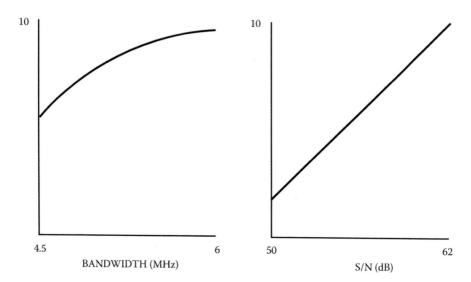

10

4.5   6

BANDWIDTH (MHz)

10

50   62

S/N (dB)

FIGURE 5.9
Attribute utility trade curve example.

The scores are filled in on the decision analysis worksheet and multiplied by the weights to calculate the weighted score. The total of the weighted scores for each candidate then determines their ranking. As a rule, at least a 10 percent difference in score is acceptable as meaningful.

Further analysis can be applied in terms of evaluating the sensitivity of the decision to changes in the value of attributes, weights, subjective estimates, and cost. Scores should be checked to see if changes in weights or scores would reverse the choice.

A trade table (Figure 5.10) can be prepared to summarize the selection results. Pertinent criteria are listed for each alternative solution. The alternatives may be described in a quantitative manner, such as high, medium, or low.

Finally, the results of the trade study are documented in the form of a report, which discusses the reasons for the selections and may include the trade tree and the trade table.

A formal system of change control must be implemented throughout the systems engineering process to prevent changes from being made without proper review and ensure approval by all concerned parties and that all parties are kept informed. Change control also ensures that all documentation is kept up to date and can help to eliminate redundant documents. Finally, change control helps to control project costs.

5.2.4 Description of System Elements

Five categories of interacting system elements can be defined: equipment (hardware), software, facilities, personnel, and procedural data. Performance,

Criteria	Cool room only. Only normal convection cooling within enclosures.	Forced cold air ventilation through rack then directly into return.	Forced cold air ventilation through rack, exhausted into the room, then returned through the normal plenum.
Cost	Lowest. Conventional central air conditioning system is used.	High. Dedicated ducting required. Separate system required to cool the room.	Moderate. Dedicated ducting required for input air.
Performance Equipment Temperature Room Temperature	Poor. 80–120° F+. 65–70° F, typical as set.	Very good. 55–70° F, typical. 65–70° F, typical as set.	Very good. 55–70° F, typical. 65–70° F, typical as set.
Control of Equipment Temperature	Poor. Hot spots will occur within enclosures.	Very good.	Very good. When the thermostat is set to provide a comfortable room temperature, the enclosure will be cool inside.
Control of Room Temperature	Good. Hot spots may still exist near power hungry equipment.	Good.	Good. If the enclosure exhaust air is comfortable for operators, the internal equipment must be cool.
Operator Comfort	Good.	Good. Separate room ventilation system required; can be set for comfort.	Good. When the thermostat is set to provide a comfortable room temperature, the enclosure will be cool inside.

FIGURE 5.10
Trade table example.

design, and test requirements must be specified and documented for equipment, components, and computer software elements of the system. It may be necessary to specify environmental and interface design requirements, which are necessary for proper functioning of system elements within a facility.

The documentation produced by the systems engineering process controls the evolutionary development of the system. Figure 5.11 illustrates the special-purpose documentation used by one organization in each step of the systems engineering process.

The requirements are formalized in written specifications. In any organization, there should be clear standards for producing specifications. This can help reduce the variability of technical content and improve product quality as a result. It is also important to make the distinction here that the product should not be overspecified to the point of describing the design or making it too costly. On the other hand, requirements should not be too general or so vague that the product would fail to meet the customer needs. In large

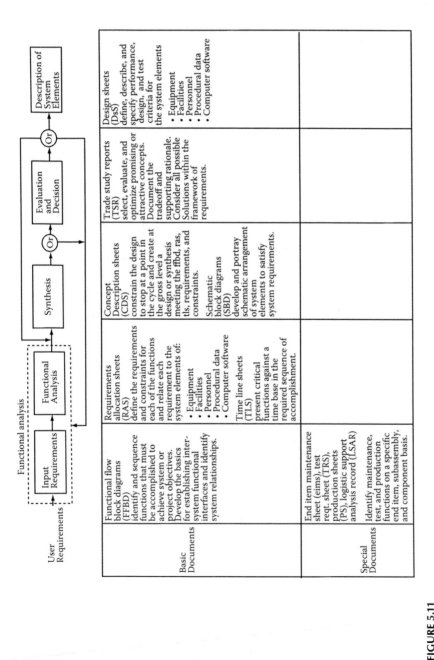

FIGURE 5.11

Basic and special-purpose documentation for system engineering. (Adapted from Hoban, F. T., and W. M. Lawbaugh, *Readings in Systems Engineering*, National Aeronautics and Space Administration, Science and Technical Information Program, Washington, DC, 1993.)

departmentalized organizations, commitment to schedules can help assure that other members of the organization can coordinate their time.

The systems engineering process does not actually design the system. The systems engineering process produces the documentation necessary to define, design, develop, and test the system. The technical integrity provided by this documentation ensures that the design requirements for the system elements reflect the functional performance requirements, that all functional performance requirements are satisfied by the combined system elements, and that such requirements are optimized with respect to system performance requirements and constraints.

5.3 Phases of a Typical System Design Project[*]

System design is carried out in a series of steps that lead to an operational unit. Appropriate research and preliminary design work is completed in the first phase of the project, the *design development* phase. It is the objective of this phase to fully delineate all requirements of the project and to identify any constraints. Based on initial concepts and information, the design requirements are modified until all concerned parties are satisfied and approval is given for the final design work to proceed. As illustrated in Figure 5.12, the first objective of this phase is to answer the following questions:

- What are the functional requirements of the product?
- What are the physical requirements of the product?
- What are the performance requirements of the product?
- Are there any constraints limiting design decisions?

A subset of the constraints question leads to the following questions:

- Will existing equipment be used?
- Is the existing equipment acceptable?
- Will this be a new facility or a renovation?
- Will this be a retrofit or upgrade to an existing system?
- Will this be a stand-alone system?

One of the systems engineer's most important contributions is the ability to identify and meet the needs of the customer and do it within the project

---

[*] Portions of this section were adapted from Whitaker, J. C., G. DeSantis, and C. R. Paulson, *Interconnecting Electronic Systems*, CRC Press, Boca Raton, FL, Chapter 1, 1992.

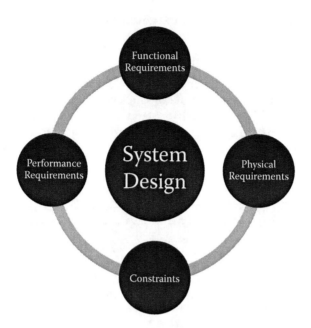

FIGURE 5.12
Typical phases of a system design project.

budget. Based on the customer's initial concepts and any subsequent equipment utilization research conducted by the systems engineer, the desired capabilities are identified as precisely as possible. Design parameters and objectives are defined and reviewed. Functional efficiency is maximized to allow operation by a minimum number of personnel. Future needs and system expansion are also investigated at this time.

After the customer approves the equipment list, preliminary system plans are drawn up for review and further development. If architectural drawings of the facility are available, they can be used as a starting point for laying out an equipment floor plan. The systems engineer uses this floor plan to be certain that adequate space is provided for present and future equipment, as well as adequate clearance for maintenance and convenient operation. Equipment identification is then added to the architect's drawings.

Documentation should include, but not be limited to, the generation of a list of major equipment including

- Equipment manufacturers, models, and prices
- Technical system functional block diagrams
- Custom item descriptions
- Rack and console elevations
- Equipment floor plans

The preliminary drawings and other supporting documents are prepared to record design decisions and to explain the design concepts to the customer. Renderings, scale models, or full-size mock-ups may also be needed to better illustrate, clarify, or test design ideas.

Ideas and concepts have to be exchanged and understood by all concerned parties. Good communication skills are essential. The bulk of the creative work is carried out in the design development phase. The physical layout—the look and feel—and the functionality of the facility will all have been decided and agreed upon by the completion of this phase of the project. If the design concepts appear feasible and the cost is within the anticipated budget, management can authorize work to proceed on the final detailed design.

5.3.1 Electronic System Design

Performance standards and specifications have to be established up front in a technical facility project. This will determine the performance level of equipment that will be acceptable for use in the system and affect the size of the budget. Signal quality, stability, reliability, and accuracy are examples of the kinds of parameters that have to be specified. Memory and processor speeds are important parameters when dealing with computer-driven products. The systems engineer has to confirm whether selected equipment conforms to the standards.

At this point it must be determined what functions each component in the system will be required to fulfill and how each will function together with other components in the system. The management and operation staff usually know what they would like the system to do and how they can best achieve it. They have probably selected equipment that they think will do the job. Being familiar with the capabilities of different equipment, the systems engineer should be able to contribute to this function-definition stage of the process. Questions that need to be answered include

- What functions must be available to the operators?
- What functions are secondary and therefore not necessary?
- What level of automation should be required to perform a function?
- How accessible should the controls be?

Over-engineering or over-design must be avoided. This serious and costly mistake can be made by engineers and company staff when planning technical system requirements. A staff member may, for example, ask for a seemingly simple feature or capability without fully understanding its complexity or the additional cost burden it may impose on a project. Other portions of the system may have to be compromised to implement the additional feature. An experienced systems engineer will be able to spot this and determine if the trade-offs, added engineering time, and cost are really justified.

When existing equipment is going to be used, it will be necessary to make an inventory list. This list will be the starting point for developing a final equipment list. Usually, confronted with a mixture of acceptable and unacceptable equipment, the systems engineer must sort out what meets current standards and what should be replaced. Then, after soliciting input from facility technical personnel, the systems engineer develops a summary of equipment needs, including future acquisitions. One of the systems engineer's most important contributions is the ability to identify and meet these needs within the facility budget.

A list of major equipment is prepared. The systems engineer selects the equipment based on experience with the products, and on customer preferences. Often, some existing equipment may be reused. A number of considerations are discussed with the facility customer to arrive at the best product selection. Some of the major points include

- Budget restrictions
- Space limitations
- Performance requirements
- Ease of operation
- Flexibility of use
- Functions and features
- Past performance history
- Manufacturer support

The goal of the systems engineer is the design of equipment to meet the functional requirements of a project efficiently and economically. Simplified block diagrams for the video, audio, control, data, and communication systems are drawn. They are discussed with the customer and presented for approval.

5.3.2 Detailed Design

With the research and preliminary design development completed, the details of the design must now be concluded. The design engineer prepares complete detailed documentation and specifications necessary for the fabrication and installation of the technical systems, including all major and minor components. Drawings must show the final configuration and the relationship of each component to other elements of the system, as well as how they will interface with other building services, such as air conditioning and electrical power. The documentation must communicate design requirements to the other design professionals, including the construction and installation contractors.

In this phase, the systems engineer develops final, detailed flow diagrams and schematics that show the interconnection of all equipment. Cable interconnection information for each type of signal is taken from the flow diagrams and recorded on the cable schedule. Cable paths are measured and timing calculations performed. Timed cable lengths (used for video and other special services) are entered in the cable schedule.

The flow diagram is a schematic drawing used to show the interconnections between all equipment that will be installed. It is different from a block diagram in that it contains much more detail. Every wire and cable must be included in the drawings. A typical flow diagram is shown in Figure 5.13.

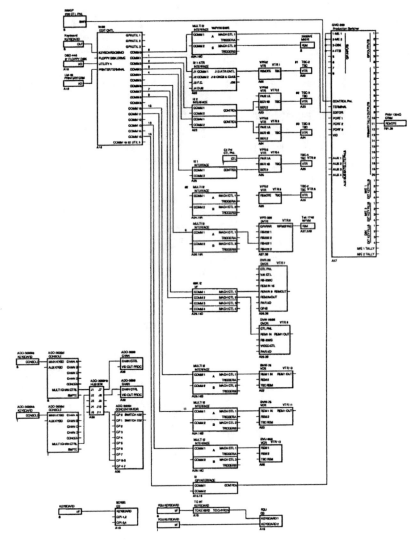

FIGURE 5.13
System control flow diagram.

The starting point for preparing a flow diagram can vary depending on the information available from the design development phase of the project and on the similarity of the project to previous projects. If a similar system has been designed in the past, the diagrams from that project can be modified to include the equipment and functionality required for the new system. New models of the equipment can be shown in place of their counterparts on the diagram, and minor wiring changes can be made to reflect the new equipment connections and changes in functional requirements. This method is efficient and easy to complete.

If the facility requirements do not fit any previously completed design, the block diagram and equipment list are used as a starting point. Essentially, the block diagram is expanded and details added to show all of the equipment and their interconnections and to show any details necessary to describe the installation and wiring completely.

An additional design feature that might be desirable for specific applications is the ability to easily disconnect a rack assembly from the system and relocate it. This would be the case if the system were to be pre-built at a systems integration facility and later moved and installed at the client's site. When this is a requirement, the interconnecting cable harnessing scheme must be well planned in advance and identified on the drawings and cable schedules.

Special custom items are defined and designed. Detailed schematics and assembly diagrams are drawn. Parts lists and specifications are finalized, and all necessary details worked out for these items. Mechanical fabrication drawings are prepared for consoles and other custom-built cabinetry.

The systems engineer provides layouts of cable runs and connections to the architect. Such detailed documentation simplifies equipment installation and facilitates future changes in the system. During preparation of final construction documents, the architect and the systems engineer can firm up the layout of the technical equipment wire ways, including access to flooring, conduits, trenches, and overhead wire ways.

Dimensioned floor plans and elevation drawings are required to show placement of equipment, lighting, electrical cable ways, duct, and conduit, as well as heating, ventilation, and air conditioning (HVAC) ducting. Requirements for special construction, electrical, lighting, HVAC, finishes, and acoustical treatments must be prepared and submitted to the architect for inclusion in the architectural drawings and specifications. This type of information, along with cooling and electrical power requirements, also must be provided to the mechanical and electrical engineering consultants (if used in the project) so that they can begin their design calculations.

Equipment heat loads are calculated and submitted to the HVAC consultant. Steps are taken when locating equipment to avoid any excessive heat buildup within the equipment enclosures, while maintaining a comfortable environment for the operators.

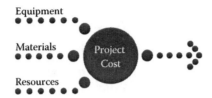

FIGURE 5.14
Primary elements of budget requirements analysis.

Electrical power loads are calculated and submitted to the electrical consultant and steps taken to provide for sufficient power and proper phase balance.

5.3.3 Budget Requirements Analysis

The need for a project may originate with customers, management, operations staff, technicians, or engineers. In any case, some sort of logical reasoning or a specific production requirement will justify the cost. On small projects, such as the addition of a single piece of equipment, a specific amount of money is required to make the purchase and cover installation costs. When the need may justify a large project, it is not always immediately apparent how much the project will cost to complete. The project has to be analyzed by dividing it up into its constituent elements. As shown in Figure 5.14, these elements include

- Equipment and parts
- Materials
- Resources (including money and time needed to complete the project)

An executive summary or capital project budget request containing a detailed breakdown of these elements can provide the information needed by management to determine the return on investment and to make an informed decision on whether or not to authorize the project.

A capital project budget request containing the minimum information might consist of the following items:

- **Project name.** Use a name that describes the result of the project, such as control room upgrade.
- **Project number (if required).** A large organization that does many projects will use a project numbering system or a budget code assigned by the accounting department.
- **Project description.** A brief description of what the project will accomplish, such as design the technical system upgrade for the renovation of production control room 2.

- **Initiation date**. The date the request will be submitted.
- **Completion date**. The date the project will be completed.
- **Justification**. The reason the project is needed.
- **Material cost breakdown**. A list of equipment, parts, and materials required for construction, fabrication, and installation of the equipment.
- **Total material cost**.
- **Labor cost breakdown**. A list of personnel required to complete the project, their hourly pay rates, the number of hours they will spend on the project, and the total cost for each.
- **Total labor cost**.
- **Total project cost**. The sum of material and labor costs.
- **Payment schedule**. Estimation of individual amounts that will have to be paid out during the course of the project and the approximate dates each will be payable.
- **Preparer name and the date prepared**.
- **Approval names, signatures, and dates approved**.

More detailed analysis, such as return on investment, can be carried out by an engineer, but financial analysis should be left to the accountants who have access to company financial data.

5.3.4 Feasibility Study and Technology Assessment

Where it is required that an attempt be made to implement new technology and where a determination must be made as to whether certain equipment can perform a desired function, it will be necessary to conduct a feasibility study. The systems engineer may be called upon to assess the state of the art to develop a new application. An executive summary or a more detailed report of evaluation test results may be required, in addition to a budget request, to help management make a decision.

5.3.4.1 Planning and Control of Scheduling and Resources

Several planning tools have been developed for planning and tracking progress toward the completion of projects and scheduling and controlling resources. The most common graphical project management tools are the Gantt chart and the *Critical Path Method* (CPM) utilizing the *Project Evaluation and Review* (PERT) technique. Computerized versions of these tools have greatly enhanced the ability of management to control large projects.

5.3.4.2 *Project Tracking and Control*

A project team member may be called upon by the project manager to report the status of the work during the course of the project. A standardized project status report form can provide consistent and complete information to the project manager. The purpose is to supply information to the project manager regarding work completed and money spent on resources and materials.

A project status report containing the minimum information might include the following items:

- Project number
- Date prepared
- Project name
- Project description
- Start date
- Completion date (the date this part of the project was completed)
- Total material cost
- Labor cost breakdown
- Preparer's name

After part or all of a project design has been approved and money allocated to build it, any changes may increase or decrease the cost. Factors that affect the cost include

- Components and material
- Resources, such as labor and special tools or construction equipment
- Costs incurred because of manufacturing or construction delays

Management will want to know about such changes and will want to control them. For this reason, a method of reporting changes to management and soliciting approval to proceed with the change may have to be instituted. The best way to do this is with a change order request or change order. A change order includes a brief description of the change, the reason for the change, and a summary of the effect it will have on costs and on the project schedule.

Management will exercise its authority to approve or reject each change based on its understanding of the cost and benefits and the perceived need for the modification of the original plan. Therefore, it is important that the systems engineer provide as much information and explanation as may be necessary to make the proposed change clear and understandable to management.

A change order form containing the minimum information might contain the following items:

- Project number
- Date prepared
- Project name
- Labor cost breakdown
- Preparer's name
- Description of the change
- Reason for the change
- Equipment and materials to be added or deleted
- Material costs or savings
- Labor costs or savings
- Total cost of this change (increase or decrease)
- Impact on the schedule

5.4 Program Management

Systems engineering is the management function that controls the total system development effort for the purpose of achieving an optimum balance of all system elements.[*] It is a process that transforms an operational need into a description of system parameters and integrates those parameters to optimize the overall system effectiveness.

Systems engineering is both a technical process and a management process. Both processes must be applied throughout a program if it is to be successful. The persons who plan and carry out a project constitute the project team. The makeup of a project team will vary depending on the size of the company and the complexity of the project. It is up to the management to provide the necessary human resources to complete the project. A typical organizational structure is shown in Figure 5.15.

5.4.1 Executive Manager

The executive manager is the person who can authorize that a project be undertaken. This person can allocate funds and delegate authority to others to accomplish the task. The ultimate responsibility for project success is in the hands of the executive manager. The executive manager assigns group responsibilities, coordinates activities between groups, and resolves

---

[*] Hoban, F. T., and W. M. Lawbaugh, *Readings in Systems Engineering Management*, National Aeronautics and Space Administration, Science and Technical Information Program, Washington, D.C., p. 9, 1993.

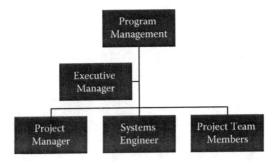

FIGURE 5.15
Elements of a program management system.

group conflicts. The executive manager establishes policy, provides broad guidelines, approves the project master plan, resolves conflicts, and ensures project compliance with commitments.

Executive management delegates the project management functions and assigns authority to qualified professionals, allocates a capital budget for the project, supports the project team, and establishes and maintains a healthy relationship with project team members.

Management has the responsibility to provide clear information and goals—up front—based on their needs and initial research. Before initiating a project, the company executive should be familiar with the daily operation of the facility and analyze how the company works, how jobs are done by the staff, and what tools are needed to accomplish the work. Some points that may need to be considered by an executive before initiating a project include the following:

- What is the current capital budget for equipment?
- Why does the staff currently use specific equipment?
- What function of the equipment is the weakest within the organization?
- What functions are needed but cannot be accomplished with current equipment?
- Is the staff satisfied with current hardware?
- Are there any reliability problems or functional weaknesses?
- What is the maintenance budget and is it expected to remain steady?
- How soon must the changes be implemented?
- What is expected from the project team?

Only after answering the appropriate questions will the executive manager be ready to bring in expert project management and engineering assistance. Unless the manager has made a systematic effort to evaluate all of the obvious points about the facility requirements, the not-so-obvious points may be overlooked. Overall requirements must be broken down into their component parts. Do not try to tackle ideas that have too many branches. Keep the planning as basic as possible. If the company executive does not make a concerted effort to investigate the needs and problems of a facility thoroughly before consulting experts, the expert advice will be shallow and incomplete, no matter how good the engineer is.

Engineers work with the information they are given. They put together plans, recommendations, budgets, schedules, purchases, hardware, and installation specifications based on the information they receive from interviewing management and staff. If the management and staff have failed to go through the planning, reflection, and refinement cycle before those interviews, the company will likely waste time and money.

5.4.2 Project Manager

Project management is the outgrowth of the need to implement large, complex projects in the shortest possible time, within the anticipated cost, and with the required performance and reliability. Project management is based on the realization that modern organizations may be so complex as to preclude effective management using traditional organizational structures and relationships. Project management can be applied to any undertaking that has a specific end objective.

The project manager must be a competent systems engineer, accountant, and manager. A systems engineer must have an understanding of analysis, simulation, modeling, and reliability and testing techniques. There must be awareness of state-of-the-art technologies and their limitations. An accountant must be aware of the financial implications of planned decisions and know how to control them. A manager must be aware that the planning and control of schedules is an important part of controlling the costs of a project and completing it on time. A manager must also have the skills necessary to communicate clearly and convincingly with subordinates and superiors to make them aware of problems and their solutions.

The project manager is the person who has the authority to carry out a project. This person has been given the legitimate right to direct the efforts of the project team members. The manager's power comes from the acceptance and respect accorded by superiors and subordinates. The project manager has the power to act and is committed to group goals.

The project manager is responsible for getting the project completed properly, on schedule, and within budget, by utilizing whatever resources are necessary to accomplish the goal in the most efficient manner. The manager provides project schedule, financial, and technical requirement direction and evaluates and reports on project performance. This requires planning, organizing, staffing, directing, and controlling all aspects of the project.

In this leadership role, the project manager is required to perform many tasks including the following:

- Assemble the project organization.
- Develop the project plan.
- Publish the project plan.
- Set measurable and attainable project objectives.
- Set attainable performance standards.
- Determine which scheduling tools (PERT, CPM, Gantt, etc.) are right for the project.
- Using the available scheduling tools, develop and coordinate the project plan, which includes the budget, resources, and the project schedule.
- Develop the project schedule.
- Develop the project budget.
- Manage the budget.
- Recruit personnel for the project.
- Select subcontractors.
- Assign work, responsibility, and authority so that team members can make maximum use of their abilities.
- Estimate, allocate, coordinate, and control project resources.
- Deal with specifications and resource needs that are unrealistic.
- Decide on the right level of administrative and computer support.
- Train project members on how to fulfill their duties and responsibilities.
- Supervise project members, giving them day-to-day instructions, guidance, and discipline as required for them to fulfill their duties and responsibilities.
- Design and implement reporting and briefing information systems or documents that respond to project needs.
- Control the project.

Some basic project management practices can improve the chances for success. Consider the following:

- Secure the necessary commitments from top management to make the project a success.
- Set up an action plan that will be easily adopted by management.
- Use a work breakdown structure that is comprehensive and easy to use.
- Establish accounting practices that help, not hinder, successful completion of the project.

- Prepare project team job descriptions properly up front to eliminate conflict later on.
- Select project team members appropriately the first time.

After the project is under way, follow these steps:

- Manage the project, but make the oversight reasonable and predictable.
- Get team members to accept and participate in the plans.
- Motivate project team members for best performance.
- Coordinate activities so they are carried out in relation to their importance with a minimum of conflict.
- Monitor and minimize interdepartmental conflicts.
- Get the most out of project meetings without wasting the team's productive time. Develop an agenda for each meeting and start on time. Conduct one piece of business at a time. Assign responsibilities where appropriate. Agree on follow-up and accountability dates. Indicate the next step for the group. Set the time and place for the next meeting. End on time.
- Spot problems and take corrective action.
- Discover the strengths and weaknesses in project team members and manage them to get the desired results.
- Help team members solve their own problems.
- Exchange information with subordinates, associates, superiors, and others about plans, progress, and problems.
- Make the best of available resources.
- Measure project performance.
- Determine, through formal and informal reports, the degree to which progress is being made.
- Determine causes of and possible ways to act upon significant deviations from planned performance.
- Take action to correct an unfavorable trend or to take advantage of an unusually favorable trend.
- Look for areas where improvements can be made.
- Develop more effective and economical methods of managing.
- Remain flexible.
- Avoid activity traps.
- Practice effective time management.

When dealing with subordinates, each person must

- Know what is to be done, preferably in terms of an end product.
- Have a clear understanding of the authority and its limits for each individual.
- Know what the relationship with other people is.
- Know what constitutes a job well done in terms of specific results.
- Know when and what is being done exceptionally well.
- Be shown concrete evidence that there are rewards for work well done and for work exceptionally well done.
- Know where and when expectations are not being met.
- Be made aware of what can and should be done to correct unsatisfactory results.
- Feel that the supervisor has an interest in each person as an individual.
- Feel that the supervisor both believes in each person and is anxious for individual success and progress.

By fostering a good relationship with associates, the manager will have less difficulty communicating with them. The fastest, most effective communication takes place among people with common points of view.

The competent project manager observes what is going on in great detail and can therefore perceive problems long before they flow through the paper system. Personal contact is faster than filling out forms. A project manager who spends much of the time roaming through the workplace usually stays on top of issues.

5.4.3 Systems Engineer

The term systems engineer means different things to different people. The systems engineer is distinguished from the engineering specialist, who is concerned with only one aspect of a well-defined engineering discipline, in that the systems engineer must be able to adapt to the requirements of almost any type of system. The systems engineer provides the employer with a wealth of experience gained from many successful approaches to technical problems developed through hands-on exposure to a variety of situations. This person is a professional with knowledge and experience, possessing skills in a specialized and learned field or fields. The systems engineer is an expert in these areas, highly trained in analyzing problems and developing solutions that satisfy management objectives. The systems engineer takes data from the overall development process and in return provides data in the form of requirements and analysis results to the process.

Education in electronics theory is a prerequisite for designing systems that employ electronic components. Being a graduate engineer, the systems engineer has the education required to design electronic systems correctly. Mathematics skill acquired in engineering school is one of the tools used by the systems engineer to formulate solutions to design problems and analyze test results. Knowledge of testing techniques and theory enables this individual to specify system components and performance and to measure the results. Drafting and writing skills are required for efficient preparation of the necessary documentation needed to communicate the design to technicians and contractors who will have to build and install the system.

A competent systems engineer has a wealth of technical information that can be used to speed up the design process and help in making cost-effective decisions. If necessary information is not at hand, the systems engineer knows where to find it. The experienced systems engineer is familiar with proper fabrication, construction, installation, and wiring techniques and can spot and correct improper work.

Training in personnel relations, a part of the engineering curriculum, helps the systems engineer communicate and negotiate professionally with subordinates and management.

Small in-house projects can be completed on an informal basis and, indeed, this is probably the normal routine when the projects are simple and uncomplicated. In a large project, however, the systems engineer's involvement usually begins with preliminary planning and continues through fabrication, implementation, and testing. The degree to which program objectives are achieved is an important measure of the systems engineer's contribution.

During the design process, the systems engineer:

- Concentrates on results and focuses work according to management objectives.
- Receives input from management and staff.
- Researches the project and develops a workable design.
- Ensures the balanced influence of all required design specialties.
- Conducts design reviews.
- Performs trade-off analyses.
- Assists in verifying system performance.
- Resolves technical problems related to the design, interface between system components, and integration of the system into any facility.

Aside from designing a system, the systems engineer has to answer any questions and resolve problems that may arise during fabrication and installation of the hardware. Quality and workmanship of the installation must be monitored. The hardware and software will have to be tested and calibrated upon completion. This, too, is the concern of the systems engineer. During the production or fabrication phase, systems engineering is concerned with verifying system capability, verifying system performance, and maintaining the system baseline.

Depending on the complexity of the new installation, the systems engineer may have to provide orientation and operating instruction to the users. During the operational support phase, system engineers

- Receive input from users.
- Evaluate proposed changes to the system.
- Facilitates the effective incorporation of changes, modifications, and updates.

Depending on the size of the project and the management organization, the systems engineer's duties will vary. In some cases, the systems engineer may have to assume the responsibilities of planning and managing smaller projects.

5.4.4 Other Project Team Members

Other key members of the project team, where building construction may be involved, include the following:

- Architect, responsible design of any structures.
- Electrical engineer, responsible for power system design if not handled by the systems engineer.
- Mechanical engineer, responsible for HVAC, plumbing and related designs.
- Structural engineer, responsible for concrete and steel structures.
- Construction contractors, responsible for executing the plans developed by the architect, mechanical engineer, and structural engineer.
- Other outside contractors, responsible for certain specialized custom items that cannot be developed or fabricated internally or by any of the other contractors.

Bibliography

Defense Systems Management, *Systems Engineering Management Guide*, Fort Belvoir, VA, Defense Systems Management College, 1983.

Delatore, J.P., E.M. Prell, and M.K. Vora, Translating customer needs into product specifications, *Quality Progress*, 22(1), January 1989.

Finkelstein, L., Systems theory, *IEE Proceedings*, Pt. A, 135(6), pp. 401–403, 1988.

Hoban, F.T., and W.M. Lawbaugh, *Readings in Systems Engineering*, National Aeronautics and Space Administrator, Washington, D.C., Science and Technical Information Program, 1993.

Shinners, S.M, *A Guide to Systems Engineering and Management*, Lexington, MA, Lexington Books, 1976.

Tuxal, J.G., *Introductory System Engineering*, New York, McGraw-Hill, 1972.

6

Concurrent Engineering*

6.1 Introduction

Concurrent Engineering (CE) is a method used to shorten the time to market for new or improved products. It can be assumed that a product will, upon reaching the marketplace, be competitive in nearly every respect—such as quality and cost, for example. But the marketplace has shown that products—even though competitive—must not be late to market because market share, and therefore profitability, will be adversely affected. Concurrent engineering is a technique that attempts to address this issue.

6.2 Overview

In the late 1970s and early 1980s, a number of proactive companies began to use what were then innovative techniques to improve their competitive position. But it was not until 1986 that a formal definition of concurrent engineering was published by the US Defense Department:

> A systematic approach to the integrated, concurrent design of products and their related processes, including manufacture, and support. This approach is intended to cause the developers, from the outset, to consider all elements of the product life cycle from concept through disposal including quality, cost, schedule, and user requirements.

This definition was printed in the Institute for Defense Analyses Report R-338 (1986). The key words are seen to be *integrated, concurrent design*, and *all elements of the product life cycle*. Implicit in this definition is the concept that in addition to input from the originators of the concept, input should

---

* This chapter is based on Long, F., "Concurrent engineering," in *The Electronics Handbook*, ed. Jerry C. Whitaker, CRC Press, Boca Raton, FL, 2005.

come from users of the product, from those who install and maintain the product, from those who manufacture and test the product, as well as from the designers of the product. Such input, as appropriate, should be in every phase of the product life cycle, even the very earliest design work.

This approach is implemented by bringing specialists from manufacturing, test, procurement, field service, etc., into the earliest design considerations. It is very different from the process so long used by industry. The earlier process, now known as the *over the wall process*, was a serial or sequential process. The product concept, formulated at a high level of company management, was then turned over to a design group. The design group completed its design effort, tossed it over the wall to manufacturing, and proceeded to an entirely new and different product design. Manufacturing tossed its product to test, and so on through the chain. The unfortunate result of this sequential process was the necessity for redesign, which happened with great regularity.

Traditional designers too frequently have limited knowledge of a manufacturing process, especially its capabilities and constraints. This may lead to a design that cannot be made economically or made in the time scheduled, or perhaps cannot be made at all. The same can be said of the specialists in the processes of test, marketing, and field service, as well as parts and material procurement. A problem in any of these areas might well require that the design be returned to the design group for rework. The same result might come from a product that cannot be repaired. The outcome is a redesign effort required to correct the deficiencies found during later processes in the product cycle. Such redesign effort is costly in both economic and time-to-market terms. Another way to view these redesign efforts is that they are not value-added. Value-added is a particularly useful parameter by which to evaluate a process or practice.

6.2.1 The Team Process

An example of a serial process is illustrated in Figure 6.1, showing that even feedback from field service might be needed in a redesign effort. When the process is illustrated in this manner, the presence of redesign can be seen to be less efficient than a process in which little or no redesign is required. A common projection of the added cost of redesign is that changes made are about 10 times more costly than correctly designing the product in the first place. If the product should be in the hands of a customer when a failure occurs, the results can be disastrous, both in direct costs to effect the repair and in possible lost sales due to a tarnished reputation.

There are two major facets of concurrent engineering that must be kept in mind at all times. The first is that a concurrent engineering process requires team effort. This involves more than the customary committee. Although the team is composed of specialists from the various activities, the team members are not there as representatives of their organizational home. They are

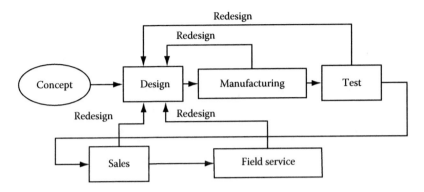

FIGURE 6.1
The serial design process.

there to cooperate in the delivery of product to the market place by contributing their expertise in the task of eliminating the redesign loops shown in Figure 6.1. Formation of the proper team is critical to the success of most CE endeavors.

The second facet is to keep in mind that concurrent engineering is information and communication intensive. There must be no barriers to complete and rapid communication among all parts of a process, even if they are located at geographically dispersed sites. If top management has access to and uses information relevant to the product or process, this same information must be available to all in the production chain, including the line workers.

An informed and knowledgeable workforce at all levels is essential, so that they may use their efforts to the greatest advantage. The most effective method to accomplish this is to form, as early as possible in the product life cycle, a team composed of knowledgeable people from all facets of the product life cycle. This team should be able to anticipate and design out most if not all possible problems before they actually occur. Figure 6.2 suggests many of the communication pathways that must be freely available to the members of the team. Other problems will surface as the project progresses. The inputs to the design process are sometimes called the "design for ...," inserting the requirement.

Top management assigns the team members and serves as the coaches for the team. They must make certain that the team members are properly trained and then allow the team to proceed with the project. The team members must be selected to have the best combination of recognized expertise in their respective fields and the best team skills. It is not always the expert in a specialty who will be the best team member. Team members, however, must have the respect for all other team members, not only for their expertise, but also for their interpersonal skills. Only then will the best products result. This concurrency of design, to include these and all other parts of the product life cycle, can measurably reduce time to market and overall investment in the product.

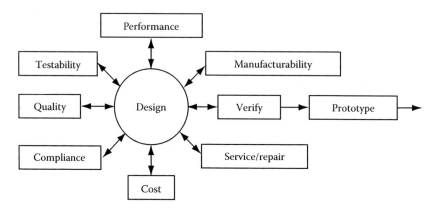

FIGURE 6.2
Concurrency of design is communication intensive.

Preventing downstream problems has another benefit in that employee morale is very likely to be enhanced. People almost universally want to take pride in what they do and produce. Few people want to support or work hard in a process or system that they believe results in a poor product. Producing a defective or shoddy product does not give them this pride. Quite often it destroys pride in workmanship and creates a disdain for the management that asks them to employ such processes or systems. The use of concurrent engineering is a very effective technique for producing a quality product in a competitive manner. Employee morale is nearly always improved as a result.

Perhaps the most important aspect of using concurrent engineering is that the design phase of a product life cycle typically takes more time and effort than the original design would have expended in the serial process. However, most organizations that have used concurrent engineering report that the overall time to market is measurably shortened because product redesign is greatly reduced or eliminated entirely. The time-worn phrase "time is money" takes on added meaning in this context.

Concurrent engineering can be thought of as an evolution rather than a revolution of the product life cycle. As such, the principles of *total quality management* (TQM) and *continuous quality improvement* (CQI), involving the ideas of robust design and reduction of variation, are not to be ignored. They continue to be important in process and product improvement. Concurrent engineering is not a type of reengineering. The concept of reengineering in today's business generally implies a massive restructuring of the organization of a process (or even a company), probably because the rate of improvement of a process using traditional TQM and CQI is deemed to be too slow or too inefficient (or both) to remain competitive. Still, the implementation of concurrent engineering does require and demand a certain and often substantial change in the way a company does business.

Concurrent engineering is as much a cultural change as it is a process change. For this reason, it is usually achieved with some difficulty. The extent of the difficulty is dependent on the willingness of people to accept change, which in turn is dependent on the commitment and sales skills of those responsible for installing the concurrent engineering culture. Although it is not usually necessary to reengineer—that is, to restructure—an entire organization to install concurrent engineering, it is also true that it cannot be installed like an overlay on top of most existing structures. Although some structural changes may be necessary, the most important change is in attitude, in culture. Yet it must also be emphasized that there is no one-size-fits-all pattern. Each organization must study itself to determine how best to install concurrent engineering. However, there are some considerations that are helpful in this study. The importance of commitment to a concurrent engineering culture from top management to line workers cannot be emphasized too strongly. A discussion of many fine ideas can be found in Salomone (1995).

6.3 The Process View of Production

If production is viewed as a process, the product life cycle becomes a seamless movement through the design, manufacture, test, sales, installation, and field maintenance activities. There is no competition within the organization for resources. The team has been charged with the entire product life cycle such that allocation of resources is seen from a holistic view rather than from a departmental or specialty view. The needs of each activity are evident to all team members. The product and process are seen as more than the sum of the parts.

The usual divisions of the process cycle can be viewed in a different way. Rather than discuss the obvious activities of manufacturability and testability and the others shown in Figure 6.2, the process can be viewed in terms of functional techniques that are used to implement the process cycle. Such a view might be as shown in Figure 6.3.

In this view, it is the functions of *quality function deployment* (QFD), *design of experiments* (DOE), and *process control* (PC) that are emphasized rather than the design, manufacturing, test, etc., activities. It is the manner in which the processes of design, manufacturing, and other elements are accomplished that is described. In this description, QFD is equated to analysis in the sense that the customers' needs and desires must be the driver in the design of today's products. Through the use of QFD, not only is the customer input heard (often referred to as the "voice of the customer"), it is translated into a process to produce the product. Thus, both initial product and process design are included in QFD in this view. It is important to note that the product and the process to produce the product are designed together, not just at the same time.

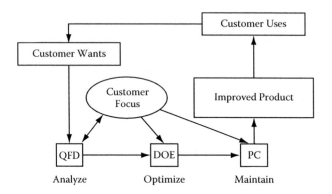

FIGURE 6.3
Functional view of the process cycle.

DOE can be used in one of two ways. First is the optimization of an existing process by removing any causes of defects and determining the best target value of the parameters. The purpose of this is to maximize the yield of a process, which frequently involves continuous quality improvement techniques. The second is the determination of a process for a new product by optimization of a proposed process before it is implemented. Today, simulation of processes is becoming increasingly important as the processes become increasingly complex. DOE, combined with simulation, is the problem-solving technique, both for running processes and proposed processes.

PC is a monitoring process to ensure that the optimized process remains an optimized process. Its primary purpose is to issue an alarm signal when a process is moving away from its optimized state. Often, this makes use of statistical methods and is then called *statistical process control* (SPC). PC is not a problem-solving technique, although some have tried to use it for that purpose. When PC signals a problem, then problem-solving techniques (possibly involving DOE) must be implemented. The following sections will expand on each of these functional aspects of a product life cycle.

6.3.1 Quality Function Deployment (QFD)

QFD begins with a determination of the customers' needs and desires. There are many ways that raw data can be gathered. Two of these are questionnaires and focus groups. Data collection is a well-developed field. The details of these techniques will not be discussed here as much has been written on the subject. It is important, however, that professionals are involved in the design of such data acquisition because of the many nuances that are inherent in such methods.

The data obtained must be translated into language that is understood by the company and its people. It is this translation that must extract the

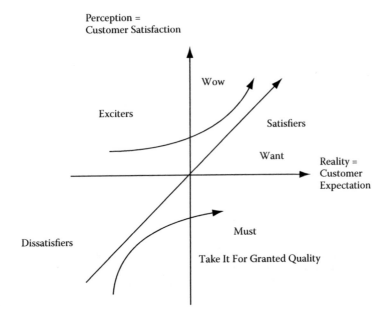

FIGURE 6.4
The Kano model.

customers' needs and wants and frame them in words that the designers, manufacturers, etc., can use in their tasks. Yet, the intent of the customers' words must not be lost. This is not always an easy task, but is a vitally important one. Another facet of this is the determination of unstated but pleasing qualities of a product that might provide a marketing edge. This idea has been sketched by the Kano model, shown in Figure 6.4

The Kano model—developed by Noriaki Kano, a Japanese professor—describes these pleasers as the "wows." The model also shows that some characteristics of a product are not even mentioned by customers or potential customers, yet they must be present or the product will be deemed unsatisfactory. The wows are not even thought of by the customers but give the product a competitive advantage. An often-cited example is the net in the trunk of an automobile that can be used to secure loose cargo such as grocery bags. Translating the customers' responses into useable items is usually accomplished by application of the *house of quality*. The house of quality is a matrix, or perhaps more accurately an augmented matrix. Two important attributes of the house of quality are (1) its capability for ranking the various inputs in terms of perceived importance, and (2) the data in the completed house that shows much of the decision making that went into the translation of customers' wants and need into useable task descriptions. The latter attribute is often called the *archival characteristic* and is especially useful when product upgrades are designed or when new products of a similar nature are designed.

Constructing the house of quality matrix begins by identifying each horizontal row of the main or *correlation matrix* with a customer input, called the *customer attribute* (CA). These CAs are entered on the left side of the matrix row. Each vertical column of the matrix is assigned an activity, called the *engineering characteristic* (EC). The ECs are entered at the top of the columns. Together, the ECs are believed by the team to be able to produce the CAs. Because a certain amount of judgment is required in determining the ECs, such TQM techniques as brainstorming are often used by the team at this time. The team must now decide the relative importance of the ECs in realizing the CAs. Again, the techniques of TQM are used. The relative importance is indicated by assigning the box, that is, the intersection of an EC with a CA a numerical value, using an agreed-upon rating scale (ECs such as blank for "unimportant" to 5 or 10 for "very important"). Each box in the matrix is then assigned a number. Note that an EC may affect more than one CA and that one CA may require more than one EC to be realized. An example of a main matrix CAs with ranking information in the boxes and a summary at the bottom is shown in Figure 6.5. In this example, the rankings of the ECs are assigned only three relative values rather than a full range of 0–9. This is frequently done to reduce the uncertainty and lost time as a result of trying to decide, for example, between a 5 or a 6 for the comparisons. Also, weighting the three levels unequally will give emphasis to the more important relationships.

Following completion of the main matrix, the augmentation portions are added. The first is usually the planning matrix that is added to the right side of the main matrix. Each new column added by the planning matrix lists items that have a relationship to one or more of the CAs but are not ECs. These customer perceptions (or attributes) might be assumed customer

		ECs										
		A	B	C	D	E	F	G	H	I	J	
CAs	1	9		3	3		1					
	2		3	1			9					
	3	3	3		1	1		3		9	1	
	4			1					1			
	5	1	1			3				1		
	6			3			1				3	
	7				9			3		1		
	8	9				3			3			
	9		3		1			1			3	
	10	3		9								
TOTAL		25	10	20	14	7	11	7	4	11	7	116
%		21	9	17	12	6	10	6	3	10	6	100

FIGURE 6.5
The house of quality main matrix.

Cas		Importance	Our Status	Improvement Needed	Competitive Advantage	Overall Weight	%	Competitor
Att 1		5	3	1.7	1.0	5.1	21	4
Att 2		3	3	1.0	1.5	4.5	19	3
Att 3		5	4	1.3	1.0	5.2	22	4
Att 4		2	3	1.0	1.0	3.0	13	2
Att 5		1	4	1.0	1.5	6.0	25	3
						23.8	100	

FIGURE 6.6
CAs and the planning matrix.

relative importance, current company status, estimated competitor's status, sales positives (wows), improvement needed, etc. Figure 6.6 shows a planning matrix with relative weights of the customer attributes for each added relationship. The range of weights for each column is arbitrary, but the relationships between columns should be such that a row total can be assigned to each CA and each row given a relative weight when compared to other CA rows.

Another item of useful information is the interaction of the ECs because some of these can be positive, reinforcing each other, whereas others can be negative: improving one can lower the effect of another. Again, the house of quality can be augmented to indicate these interactions and their relative importance. This is accomplished by adding a roof. Such a roof is shown in Figure 6.7. This is very important information to the design effort, helping to guide the designers as to where special effort might be needed in the optimization of the product.

It is very likely that the ECs used in the house of quality will need to be translated into other requirements. A useful way to do this is to use the ECs from this first house as the inputs to a second house, whose output might be cost or parts to accomplish the ECs. It is not unusual to have a sequence of several houses of quality, as shown in Figure 6.8.

The final output of a QFD utilization should be a product description and a first pass at a process description to produce the product. The product description should be traceable to the original customer inputs so that this product will be competitive in those terms. The process description should be one that will produce the product in a competitive time and cost framework. It is important to note that the QFD process, to be complete, requires input from all parts of a product cycle.

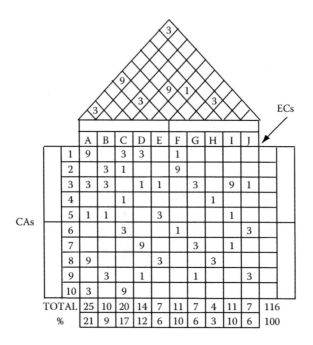

ECs

CAs

		A	B	C	D	E	F	G	H	I	J	
	1	9		3	3		1					
	2		3	1			9					
	3	3	3		1	1		3		9	1	
	4			1					1			
	5	1	1			3				1		
	6		3				1				3	
	7				9			3	1			
	8	9				3		3				
	9		3		1			1			3	
	10	3		9								
TOTAL		25	10	20	14	7	11	7	4	11	7	116
%		21	9	17	12	6	10	6	3	10	6	100

FIGURE 6.7
The house of quality with a roof.

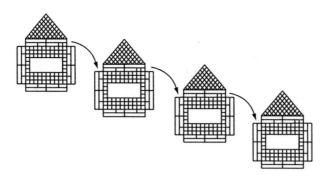

FIGURE 6.8
A series of houses of quality.

This brief look at QFD hopefully indicates the power of this tool. Initial use of this tool will most likely be more expensive than currently used methods because of the familiarization that must take place. It should not be used for small tasks, those with fewer than 10 input statements. It should also probably not be used if the number of inputs exceeds 50 because the complexity makes the relative weightings difficult to manage with a satisfactory degree of confidence.

6.3.2 Design of Experiments (DOE)

DOE is an organized procedure for identifying those parts of a process that are causing a less than satisfactory product and then optimizing the process. The process might be one already in production or it might be one that is proposed for a new product. A complex process cannot be studied effectively by using the simple technique of varying one parameter while holding all others fixed. Such a procedure, though historically taught in most science and engineering classes, ignores the possibility of interactions between parameters, a condition that very often prevails in the real world. However, if all interactions as well as primary parameters are to be tested, the number of experiments rapidly becomes large enough to be out of the question for only a few variables. DOE was developed to help reduce the number of experiments required to uncover a problem parameter. As DOE developed, it became clear that some method of determining the relative importance of some relationships of parameters needed to be found. Thus, DOE came to be dependent on statistics, and it became the task of statisticians trained in advanced methods of statistics.

Factorial analysis, the study of all parameters and all their interactions, is not time efficient for hand calculations involving more than four parameters. From this, the fractional factorial method of analysis developed, in which only the primary parameters and a selected set of their interactions are studied. Brainstorming and other similar techniques from TQM to select the interactions can help but do not remove the nagging questions about those that are left out. Were they left out because of ignorance of them by the team, or by a member of the team dominating the selection because of preconceived, perhaps, erroneous ideas?

6.3.3 Robust Design

Because design determines so much of the ultimate expenditure in producing a product, there is increasing interest in process and product designs that can survive in many different environments while maintaining their design performance with quality and production costs that are competitive in the global market. Taguchi (1993) suggested that the goalposts mentality of design, accepting values anywhere within the design specifications, is harmful to the producer and the user.

Customers generally want products to be as identical as clones. This implies that the products should have zero variability of a parameter that is important. Its value should be the target value. Any product that does not exhibit this value, even though it remains within specifications, is an indication that the customer/user is not getting from the product the very best that could be delivered.

Many designs have been made using the designer's best estimates of the worst possible combination of parameters and attempting to create a design that will still meet the specifications. For those products that demand a high degree of reliability, a defective product cannot be allowed. For these, a worst-case design is mandatory. The major difficulty in doing worst-case design is to guarantee that the combination of parameters chosen does in fact result in the worst case. The more the parameters, the more difficult it is to guarantee this. Design of experiments might be used to determine the combinations that will result in the worst case. Then, simulations should be employed to verify the choices. It can be seen that worst-case design can be quite expensive, in terms of time as well as dollars. However, most products do not require such high reliability. A method of accepting less than 100 percent yield might be cost-effective for products that experience an occasional failure without catastrophic repercussions, especially if it can be repaired easily.

Still, with the customer demanding higher quality, some form of organized design for quality and reliability must be employed. One method for doing this is to simulate the product variability using Monte Carlo statistical analysis of the designs. The specifications are set for the product, and tolerances are chosen for the components. Then, the component values of the product are made to vary according to a selected statistical distribution, such as a normal (Gaussian) or a uniform distribution. The components can be varied one at a time or multiple components can be varied at the same time. Generally, varying one component at a time is not the best technique because interactions between components can be the cause of a defective product. Again, the techniques of design of experiments can aid greatly in selecting the components to be varied. Then, the Monte Carlo analysis becomes a verification of the results obtained from DOE.

References

Aguayo, R., *Deming*, Simon and Schuster, New York, 1990.

Bhote, K., *World Class Quality*, AMACOM, New York, 1991.

Brassard, M. and D. Ritter, *The Memory Jogger II*, GOAL/QPC, Methuen, MA, 1994.

Carter, D. and B. Baker, *Concurrent Engineering*, Addison-Wesley, Reading, MA, 1992.

Cohen, L., Quality function deployment: An application perspective from Digital Equipment Corporation, *National Productivity Review*, Summer, 1988.

Department of Defense, Report R-338, Institute for Defense Analyses, Washington, D.C., 1986.

Feigenbaum, E., P. McCorduck, and H. P. Nii, *The Rise of the Expert Company*, Times Books, New York, 1988.

Miller, L., *Concurrent Engineering Design*, Society of Manufacturing Engineers, SME, Dearborn, MI, 1993.

Prybyla, J., *Optimization Doesn't Have to Hurt*, ASIC and EDA, May 1993.

Salomone, T., *What Every Engineer Should Know About Concurrent Engineering*, Marcel Dekker, New York, 1995.

Saylor, J., *TQM Field Manual*, McGraw-Hill, New York, 1995.

Shillito, M. L., and D. De Marle, *Value: Its Measurement, Design, and Management*, Wiley, New York, 1992.

Shina, S., *Concurrent Engineering and Design for Manufacture of Electronics Products*, Van Nostrand Reinhold, New York, 1991.

Taguchi, G., *Taguchi on Robust Technology Development*, ASME Press, New York, 1993.

Bibliography

Ross, P., *Taguchi Techniques for Quality Engineering*, McGraw-Hill, New York, 1988.

Shriver, B., The Concurrent Process, *Circuits Assembly*, 5(2), February 1994.

Walton, M., *The Deming Management Method*, Putnam, New York, 1986.

7

Disaster Planning and Recovery*

7.1 Introduction

Earthquakes, hurricanes, tornados, and floods come to mind when we talk about risks that can disable a facility and its staff. Risk can also be man-made. Terrorist attack, arson, utility outage, or even simple accidents that get out of hand are facts of everyday life. Facilities, especially those that convey information from government to the public during major catastrophes, must keep their people and systems from becoming disaster victims.

During earthquakes, employees sometimes have to dodge falling ceiling tiles and objects hurled off shelves, and watch computers and workstations dance in different directions. Employees sometimes become the emergency response team when a disaster disrupts the workplace and emergency responders (police, fire, ambulance) are unable to respond to all calls for help. This chapter does not have ready-to-implement plans and procedures for people, facilities, or systems. It does outline major considerations for disaster planning. It is then up to you to be the champion for emergency preparedness and planning on mission-critical projects.

Major disasters help us focus on assuring that critical systems will work during emergencies. These events expose our weaknesses. They can also be rare windows of opportunity to learn from our mistakes and make improvements for the future.

7.1.1 Developing a Disaster Plan

The centerpiece of disaster planning and recovery is a documented plan. A plan for dealing with emergencies that is based on vague concepts that are not recorded or regularly reviewed is equivalent to no plan at all. A disaster plan must be documented and agreed to by all persons that it impacts.

---

* This chapter is based on Rudman, R., "Disaster planning and recovery," in *The Electronics Handbook*, ed. J. C. Whitaker, CRC Press, Boca Raton, FL, 2005.

The principles of good documentation outlined in previous chapters for technical projects apply equally to a disaster plan. Clear, concise, and logical steps, recorded in a document that is easily available to all personnel, are critical to dealing with emergency situations. Disaster planning cannot be left to chance and memory; one is unpredictable, the other is unreliable.

Similar to a product brochure that is reviewed and updated as the product line changes, a disaster plan needs to be reviewed at regular intervals and updated when the structure and/or needs of the organization change. Readers will recall that Chapter 2 discussed the challenge of securing the necessary resources for adequate documentation of a facility. The same challenge is often experienced in getting the resources necessary to develop a comprehensive disaster plan. Unfortunately, writing a disaster plan is a project that can easily be put off until another day. The problem, of course, comes when that project is put off one day too many.

Like any good technical document, a disaster plan need not be lengthy to cover all the key points. For example, a simple wiring code table will suffice for many projects or facilities and can convey the same information as pages of detail. By the same token, a disaster plan may be reduced to a large-sized poster with specific steps that should be taken in an emergency. Copies of the poster can then be placed at key locations within the facility so that when the information is needed, it can be found.

The availability of a disaster plan is an important point, and one worth expanding upon. When an engineer, customer, or other person that requires certain technical information cannot find it in a timely manner, productivity is lost. When persons experiencing an emergency situation cannot find the information they need, damage can be done and lives can be lost. Therefore, document availability is a critical part of developing a disaster plan. It does no good to develop a comprehensive disaster plan it is sits on a shelf in a locked office during the emergency. Likewise, having a copy of the disaster plan available online from a server is great—but what if power is lost and the server is down?

Just as effective product development requires input from many different sources, a disaster plan demands input from internal stakeholders as well as outside emergency service providers. As such, development of a disaster plan should be organized and executed in the same manner as any other documentation project at a facility—a group is formed, meetings held, assignments issued, draft text developed, and so on.

Developing a review schedule for a disaster plan can minimize the chances of the plan no longer reflecting the current facility configuration or work practices. One method is to institute a review of the plan at the beginning of each calendar year. Whatever process is used to develop and maintain a disaster plan, it should be applied consistently across the organization. It makes little sense for one division of a company to be prepared for emergencies, and another division completely unaware of what could happen and how they would deal with it. In this regard, a disaster plan can be treated like any other documentation project that impacts the bottom like of the company. The difference here, of course, is that it also impacts life and safety.

7.2 Emergency Management

Figure 7.1 shows the cyclical schematic of emergency management. After an emergency has been declared, response deals with immediate issues that have to do with saving lives and property. Finding out what has broken and who has been hurt is called *damage assessment*. Damage assessment establishes the dimensions of the response that are needed. Once you know what is damaged and who is injured, trapped, or killed, you have a much better idea of how much help you need. You also know if you have what you need on hand to meet the challenge. Another aspect of response is to provide the basics of human existence: water, food, and shelter. An entire industry has sprung up to supply such preparedness cycle needs. Water is available packaged to stay drinkable for years. The same is true for food. Some emergency food is based on the old military K rations or the more modern meals ready to eat (MREs).

The second phase of the emergency cycle is called *recovery*. Recovery often begins after immediate threats to life safety have been addressed. This is difficult to determine during an earthquake or flood. A series of life threats may span hours, days, or weeks. It is not uncommon for business and government emergency management organizations to be engaged in recovery and response at the same time.

FIGURE 7.1
The response and recovery cycle.

In business, response covers these broad topics:

- Restoring lost capacity to your production and distribution capabilities
- Getting your people back to work
- Contacting customers and alternate vendors
- Contacting insurers and processing claims

A business recovery team's mission is to get the facility back to 100 percent capacity, or as close to it as possible. Forming a business recovery team should be a key action step along with conducting the business impact analysis, discussed later in this chapter. Major threats must be eliminated or stabilized so that employees or outside help can work safely. Salvage missions may be necessary to obtain critical records or equipment. Reassurance should also be a part of the mission. Employees may be reluctant to return to work or customers may be loathe to enter even if the building is not red-tagged. Red tagging is a term for how inspectors identify severely damaged buildings, an instant revocation of a building's certificate of occupancy until satisfactory repairs take place. Entry, even for salvage purposes by owners or tenants, may be prohibited or severely restricted. Recovery will take a whole new turn if the facility is marked unfit for occupancy.

The third stage of the cycle is *mitigation*. Sometimes called lessons learned, mitigation covers a wide range of activities that analyze what went right and what went wrong during response and recovery. Accurate damage assessment records, along with how resources were allocated, are key elements reviewed during mitigation. At the business level, mitigation debriefings might uncover how and why the emergency generator did not start automatically. You might find that a microwave dish was knocked out of alignment by high winds or seismic activity or that a power outage disabled critical components such as security systems. On the people side, a review of the emergency might show that no one was available to reset circuit breakers or unlock normally locked doors. There may have been times when people on shift were not fed in a timely manner. Stress, an often-overlooked factor that compounds emergencies and emergency response, may have affected performance. In the "what went right" column, you might find that certain people rose to the occasion and averted an even worse disaster, that money spent on preparations paid off, or that you were able to resume 100 percent capacity much faster than expected.

Be prepared should be the watchword of emergency managers and facilities designers. Lessons learned during mitigation should be implemented in the preparedness phase. Procedures that did not work are rewritten. Depleted supplies are replenished and augmented if indicated. New resources are ordered and stored in safe places. Training in new procedures, as well as refresher training, is done. Mock drills and exercises may be carried out. The best test of preparedness is the next actual emergency. Preparedness is the one element of the four you should adopt as a continuous process.

Emergency management, from a field incident to the state level, is based on the Incident Command System (ICS). ICS was pioneered by fire service agencies in California as an emergency management model. Its roots go further back into military command and control. ICS has been adopted by virtually every government and government public safety emergency management organization in the United States as a standard.

ICS depends on simple concepts. One is *span of control*. Span of control theory states that a manager should not directly manage more than seven people. The optimum number is five. Another ICS basic is that an emergency organization builds up in stages, from the bottom up, from the field up, in response to need. For example, a wastebasket fire in your business may call for your fire response team to activate. One trained member on the team may be all that is required to identify the source of the fire, go for a fire extinguisher on the wall, and put it out. A more serious fire may involve the entire team. The team stages up according to the situation.

7.2.1 The Planning Process

It is impossible to separate emergency planning from the facility where the plan will be put into action. Emergency planning must be integral to a functional facility. It must support the main mission and the people who must carry it out. It must work when all else fails. Designers first must obtain firm commitment and backing from top management. Commitment is always easier to get if top management has experienced first-hand a major earthquake or powerful storm. Fear is a powerful source of motivation.

Disaster planning and recovery is an art, a science, and a technology. Disaster planners have their own professional groups and certification standards. Some states provide year-round classroom training for government disaster planners. Many planners work full time for the military or in the public safety arenas of government. Others work with businesses that recognize that staying in business after a major disaster is smart business. Still others offer their skills and services as consultants to entities who need to jump-start their disaster planning process. Key points of the planning process are illustrated in Figure 7.2.

The technical support group should have responsibility or supervision over the environmental infrastructure of a critical facility. Without oversight, electronic systems are at the mercy of whoever controls that environment. Local emergencies can be triggered by preventable failures in air supply systems, roof leaks, or uncoordinated telephone, computer, or AC wiring changes. Seemingly harmless acts such as employees plugging electric heaters into the wrong AC outlet have brought down entire facilities. Successful practitioners of systems design and support must take daily emergencies into account in the overall planning process. To do otherwise risks rapid doom, if not swift unemployment.

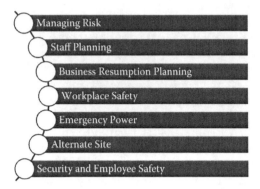

FIGURE 7.2
Steps in the disaster planning and recovery process.

7.2.2 Managing Risk

Your realistic risk list should contain specific hazards based on local conditions such as the following:

- Regional high water marks for the 100- and 150-year storms
- Regional social, political, and governmental conditions
- Regional commercial electrical power reliability
- Regional weather conditions
- Regional geography
- Regional geology

Assess specific local hazards that could be triggered by

- Threats from present or former employees who may hold grudges
- External parties who are likely to get mad at your organization
- Terrorists who may target a nearby building or area, but not specifically target your business
- Other factors that could make you an easy target
- Nearby man-made hazards
- Special on-site hazards
- Neighbors
- Construction at your facility
- Hazardous materials on the premises
- Communications links to the outside world
- Electrical power
- Other utilities
- Buried pipelines

For example, information from the Northridge (California) earthquake resulted in a rewriting of building codes for many types of structures. The Northridge quake showed that some high-rise structures thought to be earthquake safe were not. Designers should be aware that seismic building codes usually allow for safe evacuation. They do not embody design criteria to prevent major structural damage. Earthquake-safe is not earthquake-proof.

If possible, get help from emergency planning professionals when you finish your list. They can help you devise a well-written and comprehensive emergency plan. They can also help with detailed research on factors such as geology and hazardous materials. For instance, you may be located near a plant that uses hazardous materials or a rail line that may transport hazardous materials. Once there is an agreement on the major goals for operations under emergency conditions, you will have a clear direction for emergency-ready facilities planning.

7.2.3 Risk Assessment and Business Resumption Planning

Perform a realistic assessment of the risks that your list suggests. Do not overlook the obvious. If people and equipment depend on cool air, how can they continue to operate during a heat wave when your one air conditioner has malfunctioned?

What level of reliability should a designer build into a lifeline communications facility? Emergencies introduce chaos into the reliability equation. Most engineers are quite happy when a system achieves 99.9999 percent reliability. Although the glass is more than half full, four nines reliability still means 8 minutes of outage over a 1 year period. Reliability is an educated prediction based on a number of factors.

Believers in Murphy's Law ("anything that can go wrong, will go wrong") know that the 0.0001 percent outage will occur at the worst possible time. Design beyond that stage of reliability so that you can have a greater chance to cheat Murphy and stay online during major emergencies. Double or triple redundancy become realistic options when 100 percent uptime is the goal.

A facility designer must have a clear understanding of how important continued operation during and after a major catastrophe will be to its future survival. Disaster planning for facilities facing a hurricane with 125-mi/h winds may well entail boarding up the windows and leaving town for the duration. Others facilities will opt for uninterrupted operation, even in the face of nature's fury. Some will do nothing, adding to the list of victims who sap the resources of those who did prepare. The facility's mission may be critical to local government emergency management. For example, the public tunes to radio and television when disaster strikes. Government emergency managers tune in too.

A *Business Resumption Plan* (BRP) is just as important as an organization's disaster plan. Both are just as essential as having an overall strategic business plan. Some experts argue these elements should be formulated and updated in concert.

When disaster strikes, the first concern of an organization must be for the safety of its employees, customers, vendors, and visitors.

Once life safety issues have been addressed, the next step is to perform damage assessments, salvage operations, and (if needed) relocation of critical functions. The BRP is activated once these basic requirements have been met and may influence how they are met.

The focus of business resumption planning is maintaining or resuming core business activities following a disaster. The three major goals of the BRP are resumption of production and delivery, customer service and notification, and cash flow. A comprehensive BRP can accelerate recovery, saving time, money, and jobs. Similar to any insurance policy or any other investment, a properly designed BRP will cost you both in time and money. A BRP is an insurance policy, if not a major investment. The following actions are the backbone of a BRP:

- Conduct a Business Impact Analysis (BIA)
- Promote employee buy-in and participation
- Seek input starting at the lowest staff levels
- Build a recovery strategy and validation process
- Test your BRP
- Ensure a continuous update of the BRP

Most companies do not have the staff to do a proper BIA. If a BIA becomes a top management goal, retain an experienced consultant. Qualified consultants can also conduct training and design test exercises. The BIA process does work. First Interstate Bank was able to open for business the very next day after a fire shut down their Los Angeles high-rise headquarters in 1988. On January 17, 1994, the day of the Northridge earthquake, Great Western Bank headquarters suffered major structural damage. They made an almost seamless transition to their Florida operations.

7.2.4 Workplace Safety

Employers must ensure safety in the workplace at all times. Many areas have passed legislation mandating that employers identify hazards and protect their workers from them. Natural emergencies create special hazards that can maim or kill. A *moment magnitude* 7.4 earthquake can hurl heavy objects such as computer monitors and dagger-like shards from plate glass windows lethally through the air. The Richter scale is no longer used by serious seismic researchers. Moment magnitude calculates energy release based on the surface area of the planes of two adjacent rock structures (an earthquake fault) and the distance these structures will move in relation to one another. Friction across the surface of the fault holds the rocks until enough stress builds to release energy. Some of the released energy travels via low-frequency wave motion through the rock. These low-frequency waves cause the shaking and sometimes violent accelerations that occur during an earthquake. A strong foundation of day-to-day safety can lessen the impact of major emergencies. For instance, ensuring that plate glass in doors has a safety rating could avoid an accidental workplace injury.

Special and dangerous hazards often are found in the information and communications workplace. Tall equipment racks are often not secured to floors, much less secured to load-bearing walls. Preventing equipment racks from tipping over during an earthquake may avoid crippling damage to both systems and people. Bookcases and equipment storage shelves should be secured to walls. Certain objects should be tethered, rather than firmly bolted. Although securing heavy objects is mostly common sense, consult experts for special cases. Do not forget seismic-rated safety chains for heavy objects.

Computers and monitors are usually not secured to work surfaces. A sudden drop from workstation height would ruin the day's output for most computers, video monitors, and their managers. An industry has sprung up that provides innovative fasteners for computer and office equipment. Special Velcro® quick-release anchors and fasteners can support the entire weight of a personal computer or printer, even if the work surface falls over.

Bolting workstations to the floor and securing heavy equipment with properly rated fasteners can address major seismic safety issues. G forces measured in an upper story of a high-rise building during the Northridge quake were greater than 2.7 times the force of gravity; 1 g is, of course, equal to the force of Earth's gravity. An acceleration of 2 g doubles the effective force of a person or object in motion, and nullifies the effectiveness of restraints that worked fine just before the earthquake; Force = (mass) × (acceleration). Seismic accelerations cause objects to make sudden stops: 60 to zero in 1 s. A room full of unsecured workstations could do a fair imitation of a slam dance contest, even at lower accelerations. Cables can be pulled loose, monitors can implode, and delicate electronics can be smashed into scrap.

Maintaining safety standards is difficult in an organization of any size. A written safety manual that has specific practices and procedures for normal workplace hazards as well as the emergency-related hazards you identify is not only a good idea, it may lower your insurance rates. If outside workers set foot in your facility, prepare a special safety manual for contractors. Include in it installation standards, compliance with lock-out/tag-out, and emergency contact names and phone numbers.

7.2.5 Outside Plant Communications Links

Your facility may be operational, but failure of a wire, microwave, or fiber-optic communications link could be devastating. All outside plant links discussed next presuppose proper installation. For wire and fiber, this means adequate service loops (coiled slack) so that quake and wind stresses will not snap taut lines. It means that the telephone company has installed terminal equipment so it will not fall over in an earthquake or be easily flooded out. A range of backup options is available.

7.2.5.1 Outside Plant Wire

Local telephone companies still use a lot of wire. If your facility is served only by wire on telephone poles or underground in flood-prone areas, you may want what the telephone industry calls *alternate routing*. Alternate routing from your location to another *Central Office* (CO) may be very costly since the next nearest CO is rarely close. Ask to see a map of the proposed alternate route. If it is alternate only to the next block, or duplicates your telephone pole or underground risk, the advantage you gain will be minimal.

Most telephone companies can designate as an *essential service* a limited block of telephone numbers at a given location for lifeline communications. Lines so designated are usually found at hospitals and public safety headquarters. Contact your local phone company representative to see if your facility can qualify.

Even with the best of planning, outages can occur. In Spring 2009, communication cables in the San Francisco Bay Area were cut, leaving thousands of businesses, homes, and governmental agencies without Internet, land-line phone, or cell phone coverage. Banks closed their doors as they were unable to connect with their data center. ATMs and credit card readers were not able to accept transactions. Although there was no disaster, city workers were assigned to maintain a presence throughout the city—their radios available to call in for emergency services if needed.

7.2.5.2 Microwave Links

Wind and seismic activity can cause microwave dishes to go out of alignment. Earthquake-resistant towers and mounts can help prevent alignment failure, even for wind-related problems. Redundant systems should be considered part of your solution. A duplicate microwave system might lead to a false sense of security. Consider a nonmicrowave backup, such as fiber, for a primary microwave link. Smoke, heavy rain, and snowstorms can cause enough path loss to disable otherwise sound wireless systems.

Line-of-sight microwave antennas should be evaluated for potential obstructions. A microwave connection that works today may go down if a new tilt-up building blocks the signal.

7.2.5.3 Fiber Optic Links

If you are not a fiber customer today, you will be tomorrow. Telephone companies will soon be joined by other providers to seek your fiber business. You may be fortunate enough to be served by separate fiber vendors with separate fiber systems and routing to enhance reliability and uptime. Special installation techniques are essential to make sure fiber links will not be bothered by earth movement, subject to vandalism, or vulnerable to single-point failure. Single-point failure can occur in any system. Single-point failure analysis and prevention is based on simple concepts: A chain is only as strong as its weakest link, but two chains, equally strong, may have the same weak link. The lesson may be make one chain much stronger, or use three chains of a material that has different stress properties.

Fiber should be installed underground in a sturdy plastic sheath, called an *interliner*. Interliners are usually colored bright orange to make them stand out in trenches, manholes, and other places where careless digging and prodding could spell disaster. This sheath offers protection from sharp rocks or other forces that might cause a nick or break in the armor of the cable, or actually sever one or more of the bundled fibers. Cable systems that only have aerial rights-of-way on utility poles for their fiber may not prove as reliable in some areas as underground fiber. Terminal equipment for fiber should be installed in earthquake-secure equipment racks away from flooding hazards. Fiber electronics should have a minimum of two parallel DC power supplies, which are in turn paralleled with rechargeable battery backup.

7.2.5.4 Satellite

Ku- or C-band satellite is a costly but effective way to link critical communications elements. C band has an added advantage over Ku during heavy rain or snowstorms. Liquid or frozen water can disrupt Ku-band satellite transmission. A significant liability of satellite transmission for ultrareliable facilities is the possibility that a storm could cause a deep fade, even for C-band links. Another liability is short but deep semiannual periods of sun outage when a link is lost while the sun is focused directly into a receive dish. Although these periods are predictable and last for only a minute or two, there is nothing that can prevent their effect unless you have alternate service on another satellite with a different sun outage time, or terrestrial backup.

7.2.6 Emergency Power and Batteries

We are very dependent on power, and sometimes take it for granted. In the 1990s, a construction crew at San Jose International Airport severed a major power line to the terminal building. The sun had gone down, so it was very dark inside and all of the ticketing computers were down. During the construction, the backup generator had been disconnected. People were beginning to panic a bit, so one resourceful college student brought out his guitar and started playing, which helped calm the crowd. After about an hour, the power was restored.

Uninterruptable power supplies (UPS) are common in the information workplace. From small UPS units that plug into wall outlets at a personal computer workstation, to giant units that can power an entire facility, they all have one thing in common, batteries. UPS batteries have a finite lifespan. Once this lifespan is exceeded, a UPS is nothing more than an expensive doorstop. UPS batteries must be tested regularly. Allow the UPS to go online to test it. Some UPSs test themselves automatically. Some UPS applications require hours of power, whereas some only need several minutes. Governing factors are

- Availability of emergency power that can be brought online fast
- A need to keep systems alive long enough for a graceful shutdown
- Systems so critical that they can never go down

Although a UPS provides emergency power when the AC mains are dead, many are programmed with another electronic agenda: protect the devices plugged in from what the UPS thinks is bad power. Many diesel generators in emergency service are not sized for the load they have to carry, or may not have proper power factor correction. Computers and other devices with switching power supplies can distort AC power waveforms; the result: bad power.

After a UPS comes online, it should shut down after the emergency generator picks up the load and charges its batteries. If the UPS senses the poor AC power quality, the unit will stay on or cycle on and off. Its battery eventually runs down. Your best defense is to test your entire emergency power system under full load. If a UPS cycles on and off to the point that its batteries run down, you must find out why. Consult your UPS manufacturer, service provider, or service manual to see if your UPS can be adjusted to be more tolerant. Some UPS cycling cannot be avoided with engine-based emergency power, especially if heavy loads such as air conditioner compressors switch on and off line.

Technicians sometimes believe that starting an emergency generator with no equipment load is an adequate weekly test. Even a 30-min test will not get the engine up to proper operating temperature. If your generator is diesel driven, this may lead to wet stacking, cylinder glazing, and piston rings that can lose proper seating. Wet stacking occurs when a generator is run repeatedly with no load or a light load.

When the generator is asked to come online to power a full equipment load, deposits that build up during no-load tests may prevent it from developing full power. The engine will also not develop full power if its rings are not seated properly and there is significant cylinder glazing. The proper approach is to always test with the load your diesel has to carry during an emergency. If that is not possible, obtain a resistive load bank to simulate a full load for an hour or two of hard running several times per year. A really hard run should burn out accumulated carbon, reseat rings, and deglaze cylinder walls.

Fuel stored in tanks gets old, and old fuel is unreliable. Gum and varnish can form. Fuel begins to break down. Certain forms of algae can grow in diesel oil, especially at the boundary layer between fuel and the water that can accumulate at the bottom of most tanks. Fuel additives can extend the storage period and prevent algae growth. A good filtering system, and a planned program of cycling fuel through it, can extend storage life dramatically. Individual fuel chemical composition, fuel conditioners, and the age and type of storage tank all affect useful fuel life.

There are companies that will analyze your fuel. They can filter out dirt, water, and debris that can rob your engine of power. The cost of additives and fuel filtering is nominal compared to the cost of new fuel plus hazardous material disposal charges for old fuel. Older fuel tanks can spring leaks that either introduce water into the fuel or introduce you to a costly hazardous materials clean-up project. Your tank will be out of service while it is being replaced.

While you are depending on an emergency generator for your power, you would hate to see it stop. A running generator will consume fuel, crankcase oil, and possibly radiator coolant. Know your generator's crankcase oil consumption rate, so you can add oil well before the engine grinds to a screeching, nonlubricated halt. Water-cooled generators must to be checked periodically to make sure there is enough coolant in the radiator. Make sure you have enough coolant and oil to get the facility through a minimum of one week of constant duty.

Most experts recommend a generator health check every six months. Generators with engine block heaters put special stress on fittings and hoses. Vibration can loosen bolts, crack fittings, and fatigue wires and connectors. If your application is supercritical, a second generator may give you a greater margin of safety. Your generator maintenance technician should take fuel and crankcase oil samples for testing at a qualified laboratory. The fuel report will let you know if your storage conditions are acceptable. The crankcase oil report might find microscopic metal particles, early warning of a major failure.

How long will a generator last? Some engine experts say that a properly maintained diesel generator set can run in excess of 9,000 h before it would normally need to be replaced.

Mission dictates need. Need dictates reliability. If the design budget permits, a second or even third emergency generator is a realistic insurance policy. When you are designing a facility you know must never fail, consider redundant UPS wired in parallel. Consult the vendor for details on wiring needs for multiphase parallel UPS installations. During major overhauls and generator work, make sure you have a local source for reliable portable power. High-power diesel generators on wheels are common now to supply field power for events from rock concerts to movie shoots. Check the telephone directory or the web for local suppliers. If you are installing a new diesel, remember that engines over a certain size may have to be licensed by your local air quality management district and that permits must be obtained to construct and store fuel in an underground tank.

7.2.7 Air Handling Systems

Equipment crashes when it gets too hot. Clean, cool, dry, and pollutant-free air in generous quantities is critical for modern facilities. If you lease space in a high-rise, you may not have your own air system. Many building systems often have no backup, are not supervised nights and weekends, and may have uncertain maintenance histories. Your best protection is to get the exact terms for air-conditioning nailed down in your lease. You may wish to consider adding your own backup system, a costly but essential strategy if your building air supply is unreliable or has no backup. Several rental companies specialize in emergency portable industrial-strength air-conditioning. An emergency contract for heating ventilating and air-conditioning (HVAC) that can be invoked with a phone call could save you hours or even days of downtime. Consider buying a portable HVAC unit if you are protecting a critical facility.

Wherever cooling air comes from, there are times when you need to make sure the system can be forced to recirculate air within the building, temporarily becoming a closed system. Smoke or toxic fumes from a fire in the neighborhood can enter an open system. Toxic air could incapacitate your people in seconds. With some advanced warning, forcing the air system to full recirculation could avoid or forestall calamity. It could buy enough time to arrange an orderly evacuation and transition to an alternate site.

7.2.8 Water Hazards

Water in the wrong place at the wrong time can be part of a larger emergency or be its own emergency. A simple mistake such as locating a water heater where it can flood out electrical equipment can cause short circuits should the tank wear out and begin to leak. Unsecured water heaters can tear away from gas lines, possibly causing an explosion or fire in an earthquake. The water in that heater could be lost, depriving employees of a source of emergency drinking water.

The location of the water shut-off valve should be easily accessed and clearly marked. Shutting off the valve will help keep the water heater tank self-contained, thus avoiding possible contamination from the outside water supply.

Your facility may be located near a source of water that could flood your building. Many businesses are located in flood plains that see major storms once every 100 or 150 years. If you happen to be on watch at the wrong time of the century, you may wish that you had either located elsewhere, or stocked a very large supply of sandbags.

Remember to include any wet or dry pipe fire sprinkler systems as potential water hazards.

7.2.9 Alternate Sites

No matter how well you plan, something still could happen that will require you to abandon your facility for some period of time. Government emergency planners usually arrange for an alternate site for their EOCs. Communications facilities can sign mutual aid agreements. Sometimes this is the only way to access telephone lines, satellite uplink equipment, microwave, or fiber on short notice. If management shows reluctance to share, respectfully ask what they would do if their own facility is rendered useless.

7.2.10 Security

It is a fact of modern life that man-made disasters must now enter into the planning and risk assessment process. Events ranging from terrorism to poor training can cause the most mighty organization to tumble. The World Trade Center and Oklahoma City bombings are a warning to us all. Your risk assessment might even prompt you to relocate if you are too close to ground zero. Breaches in basic security have often led to serious incidents at a number of places throughout the county. Here are the basics:

- Approve visits from former employees through their former supervisors.
- Escort nonemployees in critical areas.
- Ensure that outside doors are never propped open.
- Secure roof hatches from the inside and have alarm contacts on the hatch.
- Use identification badges when employees will not know each other by sight.
- Check for legislation that may require a written safety and security plan.
- Use video security and card key systems where warranted.
- Repair fences, especially at unmanned sites.
- Install entry alarms at unattended sites; test weekly.
- Redesign to prevent unauthorized entry.
- Redesign to limit danger from outside windows.
- Plan for fire, bomb threats, hostage situations, terrorist takeovers.
- Plan a safe way to shut the facility down in case of invasion.
- Plan guard patrol schedules to be random, not predictable.
- Plan for off-site relocation and restoration of services on short notice.

7.2.11 Staff Expectations, 9-1-1, and Emergencies

A critical facility deprived of its staff will be paralyzed just as surely as if all of the equipment suddenly disappeared. Employees may experience stress if they are at work when a regional emergency strikes, and they do not know what is happening at home. The first instinct is to go home. This is often the wrong move. Blocked roads, downed bridges, and flooded tunnels are dangerous traps, especially at night. People who leave work during emergencies, especially people experiencing severe stress, often become victims. Encourage employees to prepare their homes, families, and pets for the same types of risks the workplace will face. Emergency food and water and a supply of fresh batteries are a start. Battery-powered radios and flashlights should be tested regularly. If employees or their families require special foods, prescription drugs, eyewear, oxygen, over-the-counter pharmaceuticals, sun block, or bug repellent, remind them to have an adequate supply on hand to tide them over for a lengthy emergency.

Heavy home objects such as bookcases should be secured to walls, so they will not tip over. Secure or move objects mounted on walls over beds. Make sure someone in the home knows how to shut off the water main and natural gas. The water heater tank may be used as an emergency water supply if it is not contaminated by the outside water line. An extra long hose can help for emergency fire fighting or help drain flooded areas. Suggest family hazard hunts. Educate employees on what you are doing to make the workplace safe. The same hazards that can hurt, maim, or kill in the workplace can do the same at home. Personal and company vehicles should all have emergency kits that contain basic home or business emergency supplies. Food, water, comfortable shoes, and old clothes should be added. If their families are prepared at home or on the road, employees will have added peace of mind. It may sustain them until they can get home safely.

An excellent home family preparedness measure is to identify a distant relative or friend who can be the emergency message center. Employees may be able to call a relative from work to find out their family is safe and sound. Disasters that impair telephone communications teach us that it is often possible to make and receive long distance calls when a call across the street will not get through. Business emergency planners should not overlook this hint. A location in another city, or a key customer or supplier, may make a good out-of-area emergency contact.

Television shows depicting 9-1-1 saving lives over the telephone are truly inspirational. But during a major emergency, resources normally available to 9-1-1 services, including their very telephone system, may be unavailable. Emergency experts used to tell us to be prepared to be self-sufficient at the neighborhood and business level for 72 h or more. Some now suggest a week or longer. Government will not respond to every call during a major disaster. That is a fact. The Washington State Emergency Management Division in conjunction with the California Office of Homeland Security and FEMA has developed the "Map Your Neighborhood" project to help prepare neighbors to help one another should a disaster occur. More information is available at their website: http://www.emd.wa.gov/myn.

Even experienced communications professionals sometimes forget that an overloaded telephone exchange cannot supply dial tone to all customers at once. Emergency calls will often go through if callers wait patiently for dial tone. If callers do not hear dial tone after 10 or 15 min, it is safe to assume that there is a more significant problem.

When people are trapped and professionals cannot get through, our first instinct may be to attempt a rescue. Professionals tell us that more people are injured or killed in rescue attempts during major emergencies than are actually saved. Experts in urban search and rescue (USAR) not only have the know-how to perform their work safely, but have special tools that make this work possible under impossible conditions. The jaws of life, hydraulic cutters used to free victims from wrecked automobiles, is a common USAR tool. Pneumatic jacks that look like large rubber pillows can lift heavy structural members in destroyed buildings to free trapped people.

You, as a facilities designer, may never be faced with a life-or-death decision concerning a rescue when professionals are not available. Those in the facilities you design may be faced with tough decisions. Consider that your design could make their job easier or more difficult. Also consider recommending USAR training for those responsible for online management and operations of the facility as a further means to ensure readiness.

7.3 Managing Fear

Anyone who says they are not scared while experiencing a hurricane, tornado, flood, or earthquake is either lying or foolish. Normal human reactions when an emergency hits are colored by a number of factors, including fear. As the emergency unfolds, we progress from fear of the unknown, to fear of the known. While preparedness, practice, and experience may help keep fears in check, admitting fear and the normal human response to fear can help us keep calm.

Some people prepare mentally by reviewing their behavior during personal, corporate, or natural emergencies. Then they consider how they could have been better prepared to transition from normal human reactions such as shock, denial, and panic, to desirable reactions such as grace, acceptance, and steady performance. The latter behaviors reassure those around them and encourage an effective emergency team. Grace under pressure is not a bad goal.

Another normal reaction most people experience is a rapid change of focus toward one's own personal well-being. "Am I OK?" is a very normal question at such times. Even the most altruistic people have moments during calamities when they regress. They temporarily become selfish children. Once people know they do not require immediate medical assistance, they can usually start to focus again on others and on the organization.

References

Baylus, E., *Disaster Recovery Handbook*, Chantico, New York, 1992.

Fletcher, R., *Federal Response Plan*, Federal Emergency Management Agency, Washington, D.C., 1990.

Handmer, J. and D. Parker, *Hazard Management and Emergency Planning*, James and James Science, New York, 1993.

Bibliography

Rothstein Associates, *The Rothstein Catalog on Disaster Recovery and Business Resumption Planning*, Rothstein Associates, 1993.

8

Standards and Reference Data

8.1 Introduction

Standardization usually starts within a company as a way to reduce costs associated with parts stocking, design drawings, training, and retraining of personnel. The next level might be a cooperative agreement between firms making similar equipment to use standardized dimensions, parts, and components. Competition, trade secrets, and the NIH factor ("not invented here") often generate an atmosphere that prevents such an understanding. Enter the professional engineering society, which offers a forum for discussion between users and engineers while downplaying the commercial and business aspects.

To those outside the standardization process, the wheels of progress may appear to turn very slowly. This is the inevitable result of considering all sides of an issue. Work on any standard usually begins with one or more proposals from one or more organizations. To the proponent, their submission is usually believed to be complete and sufficient to address all observed needs. The potential user of the standard, however, may have another opinion. When multiple proposals are offered to address a single need (or collection of needs), the process can become quite involved, and time-consuming.

While nearly everyone involved in standards work would like to see projects move swiftly through the process, they realize that no single organization or group has all the answers to a particular problem—or has even thought of all the questions. A great deal of creativity emerges from the competition of ideas. Finding solutions to complex problems takes time and requires participants to occasionally give up on their favored approach and agree that someone else's approach is better. It is this focus on developing the best ideas that makes the process work.

It is clear that no process involving humans is perfect, and critics can always find examples of standardization efforts that failed to meet the requirements of the user and were therefore never implemented, or implemented and then quickly faded away. This situation may be the result of rushing the process and not considering all of the sides of a particular issue.

It may also be the result of timing. Standardization work, similar to most any other product, can be adversely impacted by timing. A standard may be developed and finalized too early in the technology life cycle. If so, it may be outdated by the time it is issued because the underlying technology has continued to move forward, leaving it of little value in the marketplace. On the other hand, a standard developed too late in the process may remain unused because a proprietary solution reached the market first and has become a de facto standard.

It is difficult to address timing when it comes to standardization work. Technologists do their best to issue standards at the point the underlying technology is stable and the user base wants the product. As a practical matter, each standards organization offers its work to the marketplace. Many times they are successful; sometimes they are not. Taken on the whole, however, the work advances the state of the art, and presumably the position of the sponsoring organizations.

8.1.1 The Standards Development Organization

Membership in a Standards Development Organization (SDO) may be international, regional, national, industry, or "open source." Participation may be limited to one member/one country, one member/one company, or individualized membership. As illustrated in Figure 8.1, there are typically two types of SDOs:

- Fundamental standards organization, which sets core standards for specialized applications, generally from a clean sheet of paper.
- Applications standards organization, which develops standards used for specific market purposes. This type of group may use a mixture of fundamental standards from several SDOs and also create "glue" standards that tie different standards together for a particular application.

There are a number of considerations for various SDOs that complicate the standards development process, including the relationship of standards to regulatory bodies and intellectual property provisions. In the latter case, SDOs nearly always have clearly defined patent policies and IPR (intellectual property rights) provisions to facilitate the exchange of ideas in an open environment. The SDO, for example, may require disclosure of IPR that might be included in a draft specification, and furthermore require the participant to agree to license the IPR on Reasonable and Non-Discriminatory (RAND) terms.

Each SDO has a particular mix of document types that it develops; Figure 8.2 shows the most common ones. Generally speaking, they come under the following broad categories:

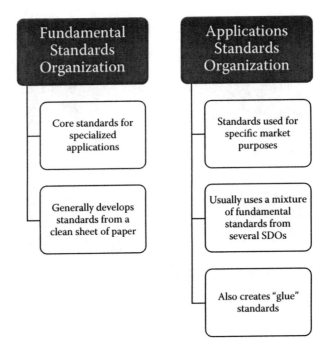

FIGURE 8.1
Basic types of SDOs.

- Standard definitions (nomenclature)
- Standard units (e.g., English or MKS/CGS)
- Test methods
- Technical specifications for interoperability
- Standard (recommended) practices

Standards are seldom available at no cost; however, there are exceptions. The number of work-hours that go into the development of a typical standard is considerable, and while the participant's time and expenses may be "donated" by their employers, the SDO has associated administrative expenses that must be recovered.

8.1.2 Professional Society Engineering Committees

The engineering groups that collate and coordinate activities that are eventually presented to standardization bodies encourage participation from all concerned parties. Meetings are often scheduled in connection with technical conferences to promote greater participation. Other necessary meetings are usually scheduled in geographical locations of the greatest activity in the field. There are usually no charges to attend the meetings. An interest in

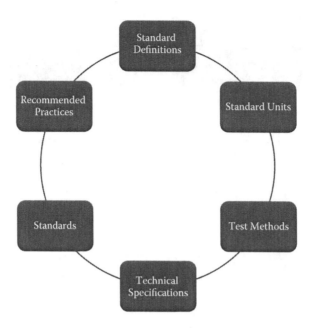

FIGURE 8.2
Common types of documents developed by an SDO.

these activities can also be served by reading the reports from these groups in the appropriate professional journals. These wheels may seem to grind exceedingly slowly at times, but the adoption of standards that may be used for 50 years or more should not be taken lightly.

8.1.3 The History of Modern Standards

Key dates in the development of modern standards are given in Figure 8.3. As shown, in 1836, the U.S. Congress authorized the Office of Weights and Measures (OWM) for the primary purpose of ensuring uniformity in custom house dealings. The Treasury Department was charged with its operation. As advancements in science and technology fueled the Industrial Revolution, it was apparent that standardization of hardware and test methods was necessary to promote commercial development and to compete successfully with the rest of the world. The Industrial Revolution in the 1830s introduced the need for interchangeable parts and hardware. Economical manufacture of transportation equipment, tools, weapons, and other machinery was possible only with mechanical standardization.

By the late 1800s, professional organizations of mechanical, electrical, chemical, and other engineers were founded with this aim in mind. The Institute of Electrical Engineers developed standards between 1890 and 1910 based on the practices of the major electrical manufacturers of the time. Such activities were not within the purview of the OWM, so there was no government involvement

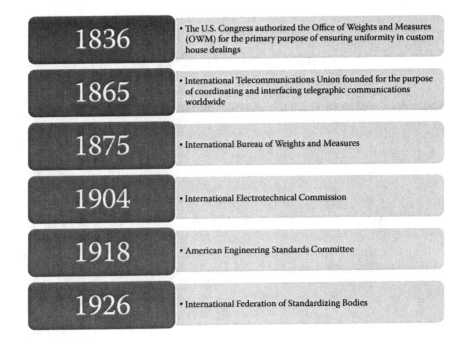

FIGURE 8.3
Timeline of key dates in standards development.

during this period. As a result of the pressures of war production in 1918, the American Engineering Standards Committee (AESC) was formed to coordinate the activities of various industry and engineering societies. This group became the American Standards Association (ASA) in 1928.

Parallel developments would occur worldwide. The International Bureau of Weights and Measures was founded in 1875, the International Electrotechnical Commission (IEC) in 1904, and the International Federation of Standardizing Bodies (ISA) in 1926. Following World War II (1946), this group was reorganized as the International Organization for Standardization (ISO), comprising the ASA and the standardizing bodies of 25 other countries.

The International Telecommunications Union (ITU) was founded in 1865 for the purpose of coordinating and interfacing telegraphic communications worldwide. Today, its member nations develop regulations and voluntary recommendations, and coordinate telecommunications development. A subgroup, the International Radio Consultative Committee (CCIR), was concerned with certain transmission standards and the compatible use of the frequency spectrum, including geostationary satellite orbit assignments.

Standardized transmission formats to allow interchange of communications over national boundaries are the purview of the ITU-R committee. Because these standards involve international treaties, negotiations are channeled through national administrations.

8.2 Principal Standards Organizations

Hundreds of SDOs exist worldwide, each targeted at a specific segment of technology and markets. In the following sections, some of the major organizations are described in brief.

8.2.1 International Organization for Standardization (ISO)

ISO is the world's largest developer and publisher of international standards. ISO is a network of the national standards institutes of more than 150 countries, one member per country, with a central Secretariat in Geneva, Switzerland, that coordinates the system.

ISO is a nongovernmental organization that forms a bridge between the public and private sectors. On the one hand, many of its member institutes are part of the governmental structure of their countries, or are mandated by their government. On the other hand, other members have their roots uniquely in the private sector, having been set up by national partnerships of industry associations. As such, ISO enables a consensus to be reached on solutions that meet both the requirements of business and the broader needs of society.

The ISO is one of three global sister organizations (ISO, IEC, ITU) that develop international standards for the world. When appropriate, these groups cooperate to ensure that international standards fit together seamlessly and complement each other. Joint committees ensure that international standards combine all relevant knowledge of experts working in related areas.

8.2.2 International Electrotechnical Commission (IEC)

The International Electrotechnical Commission is a global organization that publishes consensus-based international standards and manages conformity assessment systems for electric and electronic products, systems, and services, collectively known as *electrotechnology*. IEC publications serve as a basis for national standardization and as references when drafting international tenders and contracts. Over 10,000 experts from industry, commerce, government, test and research labs, academia, and consumer groups participate in IEC standardization work.

Founded in 1906, the IEC provides a platform to companies, industries, and governments for meeting, discussing, and developing the international standards they require. All IEC international standards are consensus based and represent the needs of key stakeholders of the nations participating in IEC work. Every member country has one vote and a say in what goes into an IEC international standard.

The IEC promotes world trade and economic growth and encourages the development of products, systems, and services that are safe, efficient, and environmentally friendly.

8.2.3 International Telecommunication Union (ITU)

The International Telecommunication Union (ITU) is the United Nations' specialized agency for information and communication technologies. ITU allocates global radio spectrum and satellite orbits, develops the technical standards that ensure that networks and technologies seamlessly interconnect, and strives to improve access to communications technologies to underserved communities worldwide.

Founded on the principle of international cooperation between governments (member states) and the private sector (sector members, associates, and academia), ITU is a global forum through which parties work toward consensus on a wide range of issues affecting the future direction of the communications industry. ITU has been based on public/private partnership since its inception. It currently has a membership of more than 190 countries and over 700 private-sector entities and academic institutions. ITU is headquartered in Geneva, Switzerland, and has 12 regional and area offices around the world. ITU membership represents a cross section of the global communications technology sector, from the world's largest manufacturers and carriers to small players working with new and emerging technologies, along with leading research and development institutions.

8.2.4 American National Standards Institute (ANSI)

The American National Standards Institute (ANSI) has served in its capacity as administrator and coordinator of the U.S. private sector voluntary standardization system for more than 90 years. Founded in 1918 by five engineering societies and three government agencies, the institute remains a private, nonprofit membership organization supported by a diverse constituency of private and public sector organizations.

ANSI coordinates policies to promote procedures, guidelines, and the consistency of standards development. Due process procedures ensure that participation is open to all persons who are materially affected by the activities without domination by a particular group. Written procedures are available to ensure that consistent methods are used for standards developments and appeals. Various levels of membership are available that support the U.S. voluntary standardization system as members of the ANSI federation.

The functions of ANSI include (1) serving as a clearinghouse on standards development and supplying standards-related publications and information, and (2) the following business development issues:

- Provides national and international standards information necessary to market products worldwide.
- Offers American National Standards that assist companies in reducing operating and purchasing costs, thereby assuring product quality and safety.
- Offers an opportunity to voice opinion through representation on numerous technical advisory groups, councils, and boards.
- Furnishes national and international recognition of standards for credibility and force in domestic commerce and world trade.
- Provides a path to influence and comment on the development of standards in the international arena.

Prospective standards must be submitted by an ANSI-accredited standards developer. There are three basic methods that may be used:

- *Accredited organization method.* This approach is most often used by associations and societies having an interest in developing standards. Participation is open to all interested parties as well as members of the association or society. The standards developer must fashion its own operating procedures, which must meet the general requirements of the ANSI procedures.
- *Accredited standards committee method.* Standing committees of directly and materially affected interests develop documents and establish consensus in support of the document. This method is most often used when a standard affects a broad range of diverse interests or where multiple associations or societies with similar interests exist. These committees are administered by a secretariat, an organization that assumes the responsibility for providing compliance with the pertinent operating procedures. The committee can develop its own operating procedures consistent with ANSI requirements, or it can adopt standard ANSI procedures.
- *Accredited canvass method.* This approach is used by smaller trade associations or societies that have documented current industry practices and desire that these standards be recognized nationally. Generally, these developers are responsible for a small number of standards. The developer identifies those who are directly and materially affected by the activity in question and conducts a letter ballot canvass of those interests to determine consensus. Developers must use standard ANSI procedures.

Note that all methods must fulfill the basic requirements of public review, voting, consideration, and disposition of all views and objections, and an appeals mechanism.

In order to maintain ANSI accreditation, standards developers are required to consistently adhere to a set of requirements or procedures known as the "ANSI Essential Requirements," which govern the consensus development process. Due process is the key to ensuring that standards are developed in an environment that is equitable, accessible, and responsive to the requirements of various stakeholders. The open and fair process ensures that all interested and affected parties have an opportunity to participate in a standard's development. It also serves and protects the public interest since standards developers accredited by ANSI must meet the institute's requirements for openness, balance, consensus, and other due process safeguards.

The introduction of new technologies or changes in the direction of industry groups or engineering societies may require a mediating body to assign responsibility for a developing standard to the proper group. The Joint Committee for Intersociety Coordination (JCIC), for example, operates under ANSI to fulfill this need.

ANSI promotes the use of U.S. standards internationally, advocates U.S. policy and technical positions in international and regional standards organizations, and encourages the adoption of international standards as national standards where they meet the needs of the user community.

8.3 Tabular Data

Whenever possible, documents should use standard values that have industry-wide acceptance. This helps to eliminate confusion on the part of readers and tends to facilitate usability outside of specific industries.

Data in Tables 8.1 to 8.8 are adapted from *The Electronics Handbook* (Whitaker, 1996).

TABLE 8.1

Common Standard Units

Name	Symbol	Quantity
ampere	A	Electric current
ampere per meter	A/m	Magnetic field strength
ampere per square meter	A/m^2	Current density
becquerel	Bg	Activity (of a radionuclide)
candela	Cd	Luminous intensity
coulomb	C	Electric charge
coulomb per kilogram	C/kg	Exposure (x and gamma rays)
coulomb per sq. meter	C/m^2	Electric flux density
cubic meter	m^3	Volume
cubic meter per kilogram	m^3/kg	Specific volume
degree Celsius	°C	Celsius temperature
farad	F	Capacitance
farad per meter	F/m	Permittivity
henry	H	Inductance
henry per meter	H/m	Permeability
hertz	Hz	Frequency
joule	J	Energy, work, quantity of heat
joule per cubic meter	J/m^3	Energy density
joule per kelvin	J/K	Heat capacity
joule per kilogram K	J/(kg·K)	Specific heat capacity
joule per mole	J/mol	Molar energy
kelvin	K	Thermodynamic temperature
kilogram	kg	Mass
kilogram per cubic meter	kg/m^3	Density, mass density
lumen	lm	Luminous flux
lux	lx	Luminance
meter	m	Length
meter per second	m/s	Speed, velocity
meter per second sq.	m/s^2	Acceleration
mole	Mol	Amount of substance
newton	N	Force
newton per meter	N/m	Surface tension
ohm	Ω	Electrical resistance
pascal	Pa	Pressure, stress
pascal second	Pa·s	Dynamic viscosity
radian	Rad	Plane angle
radian per second	rad/s	Angular velocity
radian per second square	rad/s^2	Angular acceleration
second	s	Time
siemens	S	Electrical conductance

TABLE 8.1 *(Continued)*

Common Standard Units

Name	Symbol	Quantity
square meter	m^2	Area
steradian	Sr	Solid angle
tesla	T	Magnetic flux density
volt	V	Electrical potential
volt per meter	V/m	Electric field strength
watt	W	Power, radiant flux
watt per meter kelvin	$W/(m \cdot K)$	Thermal conductivity
watt per square meter	W/m^2	Heat (power) flux density
weber	Wb	Magnetic flux

Source: Whitaker, J.C. (Ed.), *The Electronics Handbook*, CRC Press, Boca Raton, FL, 1996.

TABLE 8.2

Standard Prefixes

Multiple	Prefix	Symbol
10^{18}	exa	E
10^{15}	peta	P
10^{12}	tera	T
10^9	giga	G
10^6	mega	M
10^3	kilo	k
10^2	hecto	h
10	deka	da
10^{-1}	deci	d
10^{-2}	centi	c
10^{-3}	milli	m
10^{-6}	micro	μ
10^{-9}	nano	n
10^{-12}	pico	p
10^{-15}	femto	f
10^{-18}	atto	a

Source: Whitaker, J.C. (Ed.), *The Electronics Handbook*, CRC Press, Boca Raton, FL, 1996.

TABLE 8.3

Common Standard Units for Electrical Work

Unit	Symbol
centimeter	cm
cubic centimeter	cm³
cubic meter per second	m³/s
gigahertz	GHz
gram	G
kilohertz	kHz
kilohm	kΩ
kilojoule	kJ
kilometer	km
kilovolt	kV
kilovoltampere	kVA
kilowatt	kW
megahertz	MHz
megavolt	MV
megawatt	MW
megohm	MΩ
microampere	μA
microfarad	μF
microgram	μg
microhenry	μH
microsecond	μs
microwatt	μW
milliampere	mA
milligram	mg
millihenry	mH
millimeter	mm
millisecond	ms
millivolt	mV
milliwatt	mW
nanoampere	nA
nanofarad	nF
nanometer	nm
nanosecond	ns
nanowatt	nW
picoampere	pA
picofarad	pF
picosecond	ps
picowatt	pW

Source: Whitaker, J.C. (Ed.), *The Electronics Handbook*, CRC Press, Boca Raton, FL, 1996.

TABLE 8.4

Names and Symbols for the SI Base Units

Physical quantity	Name of SI unit	Symbol for SI unit
Length	meter	M
Mass	kilogram	Kg
Time	second	s
Electric current	ampere	A
Thermodynamic temperature	kelvin	K
Amount of substance	mole	mol
Luminous intensity	candela	cd

Source: Whitaker, J.C. (Ed.), *The Electronics Handbook*, CRC Press, Boca Raton, FL, 1996.

TABLE 8.5

Units in Use Together with the SI

Physical quantity	Name of unit	Symbol for unit	Value in SI units
Time	minute	min	60 s
Time	Hour	h	3600 s
Time	Day	d	86,400 s
Plane angle	degree	°	$(\pi/180)$ rad
Plane angle	minute	′	$(\pi/10800)$ rad
Plane angle	second	″	$(\pi/648000)$ rad
Length	ångstrom	Å	10^{-10} m
Area	Barn	b	10^{-28} m^2
Volume	liter	l, L	dm^3 = 10^{-3} m^3
Mass	tonne	t	Mg = 10^3 kg
Pressure	bar	bar	10^5 Pa = 10^5 N m^{-2}
Energy	Electronvolt[a]	eV (= e × V)	≈1.60218×10^{-19} J
Mass	unified atomic mass unit[a,b]	u (= m$_a$ (^{12}C)/12)	≈1.66054×10^{-27} kg

Source: Whitaker, J.C. (Ed.), *The Electronics Handbook*, CRC Press, Boca Raton, FL, 1996.

Note: These units are not part of the SI, but it is recognized that they will continue to be used in appropriate contexts.

[a] The values of these units in terms of the corresponding SI units are not exact as they depend on the values of the physical constants e (for the electronvolt) and N$_A$ (for the unified atomic mass unit), which are determined by experiment.

[b] The unified atomic mass unit is also sometimes called the Dalton, with the symbol Da.

TABLE 8.6

Derived Units with Special Names and Symbols

Physical quantity	Name of SI unit	Symbol for SI unit	Expression in terms of SI base units
Frequency[a]	hertz	Hz	s^{-1}
Force	newton	N	$m\ kg\ s^{-2}$
Pressure, stress	pascal	Pa	$N\ m^{-2} = m^{-1}\ kg\ s^{-2}$
Energy, work, heat	joule	J	$N\ m = m^2\ kg\ s^{-2}$
Power, radiant flux	watt	W	$J\ s^{-1} = m^2\ kg\ s^{-3}$
Electric charge	coulomb	C	$A\ s$
Electric potential, electromotive force	volt	V	$J\ C^{-1} = m^2\ kg\ s^{-3}\ A^{-1}$
Electrical resistance	ohm	Ω	$V\ A^{-1} = m^2\ kg\ s^{-3}\ A^{-2}$
Electric conductance	siemens	S	$\Omega^{-1} = m^{-2}\ kg^{-1}\ s^3\ A^2$
Electric capacitance	farad	F	$C\ V^{-1} = m^{-2}\ kg^{-1}\ s^4\ A^2$
Magnetic flux density	tesla	T	$V\ s\ m^{-2} = kg\ s^{-2}\ A^{-1}$
Magnetic flux	weber	Wb	$V\ s = M^2\ kg\ s^{-2}\ A^{-1}$
Inductance	henry	H	$V\ A^{-1}\ s = m^2\ kg\ s^{-2}\ A^{-2}$
Celsius temperature[b]	degree Celsius	°C	K
Luminous flux	lumen	lm	cd sr
Illuminance	lux	lx	$cd\ sr\ m^{-2}$
Activity (radioactive)	becquerel	Bq	s^{-1}
Absorbed done (of radiation)	gray	Gy	$J\ kg^{-1} = m^2\ s^{-2}$
Dose equivalent (dose equivalent index)	Sievert	Sv	$J\ kg^{-1} = m^2\ s^{-2}$
Phase angle	radian	rad	$1 = m\ m^{-1}$
Solid angle	sterdian	sr	$1 = m^2\ m^{-2}$

Source: Whitaker, J.C. (Ed.), *The Electronics Handbook*, CRC Press, Boca Raton, FL, 1996.

[a] For radial (circular) frequency and for angular velocity the unit rad s^{-1}, or simply s-, should be used, and this may not be simplified to Hz. The unit Hz should be used only for frequency in the sense of cycles per second.

[b] The Celsius temperature θ is defined by the equation $\theta/°C = T/K \neq 273.15$. The SI unit of Celsius temperature interval is the degree Celsius,°C, which is equal to the kelvin, K.

TABLE 8.7
The Greek Alphabet

Greek letter			Greek name	English equivalent
A	α		Alpha	a
B	β		Beta	b
Γ	γ		Gamma	g
Δ	δ		Delta	d
E	ε		Epsilon	ĕ
Z	ζ		Zeta	z
H	η		Eta	ē
Θ	θ	ϑ	Theta	th
I	ι		Iota	i
K	κ		Kappa	k
Λ	λ		Lambda	l
M	μ		Mu	m
N	ν		Nu	n
Ξ	ξ		Xi	x
O	ο		Omicron	ŏ
Π	π		Pi	p
P	ρ		Rho	r
Σ	σ	ς	Sigma	s
T	τ		Tau	t
Υ	υ		Upsilon	u
Φ	φ	ϕ	Phi	ph
X	χ		Chi	ch
Ψ	ψ		Psi	ps
Ω	ω		Omega	ō

Source: Whitaker, J.C. (Ed.), *The Electronics Handbook*, CRC Press, Boca Raton, FL, 1996.

TABLE 8.8

Constants

π Constants									
π	3.14159	26535	89793	23846	26433	83279	50288	41971	69399	37511
$1/\pi$	0.31830	98861	83790	67153	77675	26745	02872	40689	19291	48091
π^2	9.8690	44010	89358	61883	44909	99876	15113	53136	99407	24079
$\log_e \pi$	1.14472	98858	49400	17414	34273	51353	05871	16472	94812	91531
$\log_{10} \pi$	0.49714	98726	94133	85435	12682	88290	89887	36516	78324	38044
$\log_{10} \sqrt{2\pi}$	0.39908	99341	79057	52478	25035	91507	69595	02099	34102	92128

Constants involving e									
e	2.71828	18284	59045	23536	02874	71352	66249	77572	47093	69996
$1/e$	0.36787	94411	71442	32159	55237	70161	46086	74458	11131	03177
e^2	7.38905	60989	30650	22723	04274	60575	00781	31803	15570	55185
$M = \log_{10} e$	0.43429	44819	03251	82765	11289	18916	60508	22943	97005	80367
$1/M = \log_e 10$	2.30258	50929	94045	68401	79914	54684	36420	76011	01488	62877
$\log_{10} M$	9.63778	43113	00536	78912	29674	98645	-10			

Numerical constants									
$\sqrt{2}$	1.41421	35623	73095	04880	16887	24209	69807	85696	71875	37695
$\sqrt[3]{2}$	1.25992	10498	94873	16476	72106	07278	22835	05702	51464	70151
$\log_e 2$	0.69314	71805	59945	30941	72321	21458	17656	80755	00134	36026
$\log_{10} 2$	0.30102	99956	63981	19521	37388	94724	49302	67881	89881	46211
$\sqrt{3}$	1.73205	08075	68877	29352	74463	41505	87236	69428	05253	81039
$\sqrt[3]{3}$	1.44224	95703	07408	38232	16383	10780	10958	83918	69253	49935
$\log_e 3$	1.09861	22886	68109	69139	52452	36922	52570	46474	90557	82275
$\log_{10} 3$	0.47712	12547	19662	43729	50279	03255	11530	92001	28864	19070

Source: Whitaker, J.C. (Ed.), *The Electronics Handbook*, CRC Press, Boca Raton, FL, 1996.

References

Whitaker, Jerry C. (Ed.), *The Electronics Handbook*, CRC Press, Boca Raton, FL, 1996.

Bibliography

"About ISO," International Organization for Standardization, Geneva, Switzerland, http://www.iso.org/iso/about.htm.

"About the IEC: Vision and Mission," International Electrotechnical Commission, Geneva, Switzerland, http://www.iec.ch/about/.

"About the ITU," International Telecommunication Union, Geneva, Switzerland, http://www.itu.int/en/about/Pages/default.aspx.

"An Introduction to ANSI," American National Standards Institute, Washington, D.C., http://ansi.org/about_ansi/introduction/introduction.aspx?menuid=1.

Whitaker, J.C., and K.B. Benson (Eds.), *Standard Handbook of Video and Television Engineering*, New York, McGraw-Hill, 2000.

9

Document Templates

9.1 Introduction

While the documentation requirements of a given company or organization may be unique, a starting point is useful. As such, the following document templates are provided to help readers develop their own templates. It is often useful for a company to draft and standardize on a set of templates, available to all document developers as a way of both speeding the process of preparing documentation and ensuring a consistent look and feel to the documents produced by the organization.

The following templates are provided in this chapter:

- Style guide
- Technical document
- Document log
- Meeting sign-in sheet
- Draft agenda
- Meeting minutes
- Work plan
- Change request

The templates presented here are based on the guidance given in previous chapters. They are intended as a starting point. The final documents you develop will be specialized to your organization, reflecting specific needs. Most writers would agree, however, that having a starting point is preferable to a blank piece of paper.

Developing a template for the types of documents listed earlier is typically an iterative process, as illustrated in Figure 9.1. In the usual case, an individual or small group produces a draft template, which is then reviewed by a representative group of the intended users. Based on that review, revisions may be developed, which ultimately lead to the completed, published document.

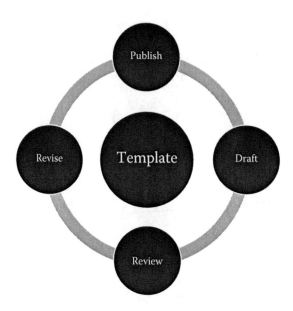

FIGURE 9.1
Cycle for developing a document template.

In an ideal case, the template will serve the needs of the organization for many years without modification. Other influences, however, may make revision of the template necessary. These include, but are not limited to

- Implementation of a new look and feel to the documentation produced by the group
- The integration of various document styles across the organization into a single look and feel
- Changing needs on the part of the typical reader

The last point deserves some additional discussion. The needs of a reader or end user tend to change over time. For example, in the early days of consumer-oriented computer software programs, extensive documentation was typically provided with the product. The documentation explained the basic concepts of the software and provided extensive reference data on syntax and other operational features. However, as software became more sophisticated, and more intuitive, the need for detailed documentation lessened. Currently, very little documentation is typically provided to users when they purchase off-the-shelf software.

The changes in software documentation illustrate the need for the documentation produced by an organization to track the real needs of the user. The enormous complexity of some modern products and systems adds another level of complexity to the documentation question. For a complex product, deciding the level of technical detail to provide can quickly become a slippery slope. The choices often boil down to one of two options:

1. Cover the subject in a superficial manner, focusing on general operation and features.
2. Cover the subject manner comprehensively, which can lead to a huge document (hundreds to thousands of pages).

As the first step in developing documentation, the template sets the architecture for the work that follows. As such, each template should be developed carefully, keeping in mind the needs of all stakeholders.

Template 1   Style Guide

Style Guide for Technical Documents

This style guide is intended to assist authors and editors of technical documents in the production of works that will ultimately become reports, brochures, and other informational documents. The guidelines given here are intended to make the process of generating documents easier for all of those involved in the effort. It is understood that not all guidelines are appropriate for all situations. Still, for the sake of consistency across all documents, this style guide should be followed to the extent possible. Consistency throughout a document ensures a professional-looking finished product, and more importantly, minimizes the potential for confusion on the part of the reader.

1 Definition of Documents

Technical documents take one of the following general forms:

- **Technical Report**—A document that states basic specifications or criteria that are necessary for effective implementation and interoperability.
- **Brochure**—A document that states specifications and operating features and capabilities of current products; may also include case histories.
- **Informational Document**—A document that incorporates user-focused information regarding product implementations.

Additional document types may be developed from time to time to address specific needs of the company.

2 General Structure of Documents

To maintain consistency across technical documents, the following guidelines should be followed in the creation of new documents. Although each type of document, and indeed each subject matter, may require special treatment, the global outline given below should be followed to the extent practical.

(a) **Cover Page.** Includes the title of document, document number, date, and revision history (if appropriate). This is always the first page of the document.

(b) **Company Page.** Includes the company mission statement, legal text, and revision history. This is always page 2 of the document.

(c) **Table of Contents.** Includes first-, second-, and third-level heads (i.e., 1, 1.1, 1.1.1), and other levels as appropriate.

(d) **Section 1—Scope.** Includes purpose, organization, and constraints as individual subheads (as needed).

(e) **Section 2—References (or Bibliography).** For references, includes normative references, informative references, and other references (as appropriate) as individual subheads. The following text should be included:

> At the time of publication, the editions indicated were valid. All referenced documents are subject to revision, and users of this document are encouraged to investigate the possibility of applying the most recent edition of the referenced document.

(f) **Section 3—Definitions and Terms.** Includes acronyms and abbreviations, global terms, syntax, and other definitions (as required) as individual subheads.

(g) **Section 4—Overview.** Provides an executive summary of the subject matter described by the document.

(h) Other sections and annexes as required.

(i) Running header as follows:

<company> <document number> (left)

<title of document> (center)

date (right)

medium weight (.75 point) rule running the column width

(j) Page numbers centered at the bottom of the page.

2.1 Fonts

In a technical document, the fewer fonts, the better. The basic recommended font set is as follows:

(a) **Body text**. Use a 12 point serif face for best readability; *Times New Roman* is recommended.

(b) **Section headings.** Use a sans serif face; *Arial* is recommended (or optionally Helvetica). Use the following sizes:

First-level head = 12 point bold

Second-level head = 12 point

Third-level head = 11 point bold

Fourth-level head = 11 point

Fifth-level head = 10 point bold

Sixth level and below = 10 point

(c) **Syntax.** Use a 9 point sans serif face within body text to indicate commands and/or code words; *Arial* is recommended (or optionally Helvetica).

(d) **Equations.** For all text other than Greek letters, use a 12 point serif face; *Times New Roman* is recommended. For Greek characters, use *Symbol*.

(e) **Hyperlinks.** Use the base font of the paragraph tag with blue characters and blue underline. When the underscore character is used in the URL, set the character to be raised by 1 point. The hyperlinks should be functional. Avoid including hyperlinks to documents or websites that are likely to change over time.

2.2 Page Layout

The recommended page layout given in this section assumes a letter-sized page. It is understood that for some applications, notably a brochure, the size may vary.

(a) **Margins**. Set the page margins for 8.5 by 11 inch paper and allow 1 inch margins on the top, bottom, and sides.

(b) **Justification and indentation**. All body text should be justified with the first line indented. Bulleted or numbered lists are usually indented and set justified.

(c) **Capitalization and italics**. All headings should be set in "title case," except first-level heads, which are in all capitals. Within the body text, capitalization (aside from the standard "sentence case") should be used with caution. Many words are capitalized by authors when there is no need to do so. When a word has a particular special meaning different from the U.S. English meaning, that meaning should be stated, and the word or phrase capitalized when used. To call attention to a term or phrase, consider using italics instead.

(d) **Table/figure numbering**. The numbering of tables and figures in the document should follow the section numbers in which the elements appear. For example, the first table in Section 5 of a document would be "Table 5.1." For tables and figures in an annex, include the annex letter followed sequentially by the number; that is, Figure A4.1.

2.3 Figures

The range of figures required for a technical document is substantial. Still, consider the following general guidelines:

(a) **Font**. Use a sans serif font such as Arial.

(b) **Finished size**. Size the illustration to be readable at the size of the printed page (8.5 by 11). For complex diagrams, consider placing the artwork at a 90° angle to provide for a larger finished illustration.

(c) **Page layout**. Artwork should be embedded in the document as close to the reference in text as practical, usually after a paragraph break. Let the artwork float centered (no box rule).

(d) **Cutline**. Place the figure cutline below the artwork with the figure number boldface (e.g., "**Figure 5.1** Descriptive text here.")

(e) **File formats**. For imported images, EPS, JPG, PNG, PDF, and TIF files are recommended. For professional printing, any line art must be scanned at 600 dpi or greater, preferably at the finished physical size.

(f) **Photographs**. Black and white photographs should be scanned at 300 dpi or greater, preferably at the finished size. Scanned black and white images should have a minimum highlight dot of 8 percent and a maximum shadow dot of 90 percent.

(g) **Screen captures**. If artwork consists of computer-screen captures, use a screen capture application capable of saving screen images at the proper resolution for printing.

(h) **Authoring program**. Authors and editors are free to develop their drawings on whatever program they are comfortable with. Still, consider using a professional drawing package, such as Adobe Illustrator or Microsoft Visio, to create drawings. Such programs offer features and time-saving elements not available in more general-purpose applications.

2.4 Tables

Providing specific guidelines for table construction is difficult because of the wide range of information required to be presented in tabular form. However, the following general guidelines are recommended.

(a) **Ruling.** Use a medium-weight (0.75 point) rule around the table, and thin rules (0.25 point) within the table to define the cells, except for the heading row (or column), which should be separated from the contents by a 0.5 point line.

(b) **Cell text.** Set the cell text left (using a sans serif font, as described previously) and use tabs (or indentation) as necessary to establish a tiered level of importance or to describe a specific structure of data or syntax.

(c) **Cutline.** Place the table cutline above the table itself.

(d) **Placement.** Place the table as close as possible to the body text that refers to it, usually after a paragraph break. Avoid tables that break across pages.

(e) **Font.** For cell text, use 9 point sans serif type; except in the case of semantic element names, which are usually 9 point bold type. For heading text, use 9 point bold sans serif type. Arial is recommended, or optionally Helvetica.

(f) **Syntax.** The following rules should be used with 'for' and 'if' statements in tables:

- One space after 'for', 'if', semicolon, and between ')' and '{'
- No space on either side of ' = ', ' = =', '<', or '>' except when next to syntax names or text
- No space between <syntax> and open-close-parenthesis; for example, 'descriptor()'
- No space between empty parenthesis; for example, '()'
- One space between <syntax> and an opening bracket; e.g., "foo_ descriptor() {'
- Operators 'i', 'j', and 'k' should be lowercase when used in a table

2.5 Special Cases

(a) **Hex numbering.** When a bit field is greater than16 bits long, insert a space after the hex characters representing the four sets of four bits in the first 16-bit hex field and after each subsequent set of 16 bits. This will break the presentation to improve readability; for example, '0x4454 4731'. Note that appropriate text should be used to indicate a range of hex values; that is, '0x01' through '0x11'.

(b) **Schema namespace.** The path to schema available on the company website should adhere to the following structure, adding additional elements if needed:

http://www.domain.com/XMLSchemas/≤application≥/≤year≥/
≤major version≥.≤minor version≥/≤schema-class≥/

For example: http://www.ourcompany.com/XMLSchemas/
mh/2009/1.0/genre-cs/

3 General Considerations

To facilitate interchangeability of documents, Microsoft Word is the format of choice for document creation. Final preparation of the document for publication will typically be done by staff and output in the latest version of Adobe Acrobat. Standard security will be set by staff to prevent the published documents from being changed. No other security restrictions are placed on published documents unless so directed by the project leader.

3.1 Numbering

Listings in the body text can be numbered, unnumbered, or bulleted. Punctuation should be consistent throughout a listing and follow proper grammar. The first word of each item should be capitalized. If a listing item is not a complete sentence, no punctuation is used. All items within a group should be complete sentences or incomplete sentences. A numbered list can be useful in a document, particularly if the text refers to specific items in the list. For a bullet list, the following hierarchy is recommended:

- Bullet list
  (1) Second-level list
      (a) Third-level list

3.2 Trademarks

Trademarks may be acknowledged in text in one of two ways:

- Include the registered trademark symbol (®) and an asterisk in the text. Add the footnote, "*Registered trademark of company, city, state."
- Place the registration information in parentheses in the text along with the trademark symbol.

Capitalize subsequent mentions of a trademarked name. It is not necessary to add the registration symbol to subsequent mentions.

3.3 Citing References

The general form for citing normative and informative references is as follows:

[1] (organization name or author name): (name of document; if a published book use italics, if not use quotation marks), (name of publication and editor/author, if applicable), (publisher), (city), (state), (volume number, if applicable; i.e., vol. 4), (series number, if applicable; i.e., no. 1), (page number or range, if applicable; i.e., pp. 10–20), (date).

For citing a website reference:

> [1] (organization website name): (title of web page—if given—using quotation marks), (name of author—if given), (URL in the form <http//www.[URL and path as appropriate].extension>), (date).

References are typically numbered in the order in which they appear in the document. It is understood, however, that this practice may not always be practical, especially for documents that have been revised one or more times.

Within the body of the document, it is suggested that the citation use the document name/number followed the appropriate bracketed reference number. For example: "Constraints on the transport stream described in ATSC A/53 Part [1] shall apply." It is further recommended that the revision letter, date, or number of the reference not be included in the text, but rather in the reference listing. In this way, a new version of an existing document can be referenced simply by updating the references (and not every use in the text).

3.4 Document Numbering

The numbering scheme used by a particular group may vary; however, consistency is essential to permit efficient document access by contributors and readers. The following basic numbering system is recommended:

$$xx\text{-}yyyrz\text{-}Name.ext$$

where:

xx	=	the group designation as a two-letter identifier (e.g., manufacturing = MN).
yyy	=	the document number
z	=	the revision number (numeric), beginning with '0' as the first-release.
$Name$	=	brief descriptive text (Title Case) of the file using the hyphen character to separate words (no spaces). In the case of agendas and minutes, "name" would be, specifically: "Agenda-yyyy-mm-dd" and "Minutes-yyyy-mm-dd", respectively.
ext	=	the document extension ("doc" for Microsoft Word documents, "pdf" for Adobe Acrobat, etc.)

The revision number should be incremented for each new version of the document.

4 Exception Policy

The guidelines given in this style guide should be used for all document types. It is understood that in some cases it may serve the best interests of the company to deviate from certain formatting specifications. Such issues should be addressed to the project manager for a decision.

Template 2a   Technical Document—Cover Page

Document Number

Date

Document Title
Subtitle (or product name)

Company Name

Address

City, State, zip code

Website

Template 2b   Technical Document—Page 2

The company information goes here, including the mission statement.

Legal disclaimers go here. In addition, copyright notices are included.

Revision History

Version	Approval Date	Notes

Template 2c   Technical Document—Table of Contents

Table of Contents

Index of Tables and Figures

Template 2d   Technical Document—Body Text

Document Title

1 Scope
The scope goes here.

1.1 Introduction and Background
If appropriate.

1.2 Organization
This document is organized as follows:

- Section 1—Outlines the scope of this document and provides a general introduction.
- Section 2—Lists references and applicable documents.
- Section 3—Provides a definition of terms, acronyms, and abbreviations for this document.
- Section 4—System overview
- Section 5—System specifications
- Annex A—Title

2 References
At the time of publication, the editions indicated were valid. All referenced documents are subject to revision, and users of this document are encouraged to investigate the possibility of applying the most recent edition of the referenced document.

2.1 Normative References
The following documents, in whole or in part, as referenced in this document, contain specific provisions that are to be followed strictly in order to implement a provision of this document.

[1] IEEE/ASTM: "Use of the International Systems of Units (SI): The Modern Metric System," Doc. SI 10-2002, Institute of Electrical and Electronics Engineers, New York, N.Y.

[2] Normative reference.

[3] Normative reference.

2.2 Informative References

The following documents contain information that may be helpful in applying this document.

[4] Informative reference.

[5] Informative reference.

3 Definition of Terms

With respect to definition of terms, abbreviations, and units, the practice of the Institute of Electrical and Electronics Engineers (IEEE) as outlined in the Institute's published standards [1] shall be used. Where an abbreviation is not covered by IEEE practice or industry practice differs from IEEE practice, the abbreviation in question will be described in Section 3.3 of this document.

3.1 Treatment of Syntactic Elements

This document contains symbolic references to syntactic elements used transport coding subsystems. These references are typographically distinguished by the use of a different font (e.g., `restricted`), may contain the underscore character (e.g., `sequence _ end _ code`) and may consist of character strings that are not English words (e.g., `dynrng`).

3.2 Acronyms and Abbreviation

The following acronyms and abbreviations are used within this document.

ABC   Description
XYW   Description

3.3 Terms

The following terms are used within this document.

reserved—An element that is set aside for use by a future document.
term—Definition.
term—Definition.

4 System Overview

This section includes a description of the system architecture, as appropriate. You may want to include one more or more block diagrams illustrating the interconnection of major elements.

5 System Specifications

The normative specifications for the document are detailed here and in other sections as needed. For an informative document, implementation guidance or background information important to the user is included in this section. Additional subsections are added as needed.

A sample table (Table 5.1) is given below.

Table 5.1 Sample Bit Stream Syntax

Syntax	No. of Bits	Format
component_list_descriptor() {		
descriptor_tag	8	uimsbf
descriptor_length	8	uimsbf
alternate	1	bslbf
component_count	7	uimsbf
for (i = 0; i<component_count; i++) {		
stream_type	8	uimsbf
format_identifier	32	uimsbf
length_of_details	8	uimsbf
stream_info_details()	var	
}		
}		

syntax—Description.

syntax—Description.

syntax—Description.

You may also want to utilize informative notes, such as the example below:

> *Informative note*: The values for each defined stream _ type with values less than 0xC4 are found in the Code Point Registry, which coordinates such values among cooperating standards development organizations.

5.1 Additional Subsections as Needed

Appropriate text here.

Template 2e   Technical Document—Annex

Annex A: Title

A1 Introduction
Introductory text, including the scope of the annex.

A2 References
The references for the annex, if used, should be detailed here. Optionally, the reader may be referred to the References section in the main document.

A3 Terms and Definitions
The terms and definitions for the annex, if used, should be detailed here. Optionally, the reader may be referred to the Terms and Definitions section in the main document.

A4 Overview
An overview of the system described in the annex.

A5 System Details
Appropriate text, tables, and illustrations.

Template 3   Document Log

Document No.	Date	Author/Editor	Title
001			
002			
003			
004			
005			
006			
007			
008			
009			
010			
011			
012			
013			
014			
015			
016			
017			
018			
019			
020			
021			
022			
023			
024			
025			
026			
027			
028			
029			

Template 4   Sign-in Sheet

Committee Name:	Committee Chair:	Date/Time:	Meeting Location:

Confidentiality: This is a meeting is covered by the policies of the company. Attendees agree that public disclosure pertaining to this meeting emanate only from, and be authorized only by, the chairperson of this committee. By remaining in attendance at this meeting, you, on behalf of yourself and your organization, agree to refrain from either directly or indirectly engaging in publicity pertaining to the business transacted at this meeting unless authorized by the chairperson.

Adherence to Company Rules: As a participant in this meeting you hereby agree, on behalf of yourself and your organization, to abide by all company rules including its Confidentiality Policy and Patent Policy.

Attendees unwilling to agree to these terms must leave the meeting.

Attendee	Affiliation	E-mail Address	Phone Number

Template 5   Draft Agenda

Draft Agenda:
Group Name

Meeting day, time, and location
Phone bridge login information
Web resources link and login information

Confidentiality: This is a meeting is covered by the policies of the company. Attendees agree that public disclosure pertaining to this meeting emanate only from, and be authorized only by, the chairperson of the committee. By remaining in attendance at this meeting, you, on behalf of yourself and your organization, agree to refrain from either directly or indirectly engaging in publicity pertaining to the business transacted at this meeting unless authorized by the chairperson.

Adherence to Company Rules: As a participant in this meeting you hereby agree, on behalf of yourself and your organization, to abide by all company rules including its Confidentiality Policy and Patent Policy.

Attendees unwilling to agree to these terms must leave the meeting.

1 **Welcome and Roll Call**
2 **Consideration of Draft Agenda**
3 **Approval of Minutes from Previous Meeting**
4 **Introductory Remarks from the Chairperson**
5 **Review of Action Items from the Previous Meeting**
6 **Specific Activities as Required**
7 **Other Business**
8 **Next Meeting Date and Location**
9 **Action Item Review**
10 **Adjournment**

Template 6   Meeting Minutes

Draft Minutes:

Group Name

Meeting day, time, and location
Phone bridge login information
Web resources link and login information

Confidentiality: This is a meeting is covered by the policies of the company. Attendees agree that public disclosure pertaining to this meeting emanate only from, and be authorized only by, the chairperson of the committee. By remaining in attendance at this meeting, you, on behalf of yourself and your organization, agree to refrain from either directly or indirectly engaging in publicity pertaining to the business transacted at this meeting unless authorized by the chairperson.

Adherence to Company Rules: As a participant in this meeting you hereby agree, on behalf of yourself and your organization, to abide by all company rules including its Confidentiality Policy and Patent Policy.

Attendees unwilling to agree to these terms must leave the meeting.

1 Welcome, Introductions, and Determination of Quorum

This section will include reference to the attendance list (typically as an attachment).

2 Approval of Proposed Agenda

An agenda should be prepared for each meeting by the chairperson. Any modifications to the printed (or posted) agenda are listed here.

3 Approval of the Draft Minutes of the Previous Meeting

Any changes or corrections offered by committee members to the minutes from the previous meeting are detailed in this section.

4 Opening Remarks

A brief summary of the opening remarks of the chairperson.

5 Action Items from the Previous Meeting

The chairperson will review the actions items from the previous meeting and a determination will be made by the group regarding the status of the items listed—typically "completed," "carried forward," or "overtaken by events."

6 Status Reports from Standing Group(s) or Ad Hoc Group(s)

This section will summarize the reports given at the meeting. If written reports or visual presentations are given, these should be included as attachments of as links to documents online.

7 New Business

A summary of the actions taken are recorded in this section. Specific points brought up during the discussion are also recorded, as appropriate.

8 Other Business

Any other business before the group will be entertained by the chairperson and recorded in this section.

9 Action Items

The action items developed at the meeting will be listed in this section.

10 Schedule of Next Meeting

The next meeting of the group will be decided or confirmed. All groups are encouraged to plan their meetings as far in advance as practical to facilitate efficient travel arrangements for attendees.

11 Adjournment

The time the meeting was adjourned.

Template 7   Work Plan

Group Name

Work Plan and Schedule

1 Scope of Work
This section describes the scope of work being undertaken by the group in general, and this project in particular.

2 Objective of the Work
This section describes the expected outcome of the work of this group.

3 Work Plan Milestones
The work plan and schedule is a key element in keeping a project on time. The following table should be modified as required, based on the project or task.

Item Number	Task	Completion Date	Comments
1	Initial planning meeting		
2	Define requirements		
3	Develop requests for proposals		
4	Review proposals and arrive at action plan		
5	Evaluation process		
6	Begin preparation of technical document		
7	Approval of document by group		
8	Project document submitted to parent group		
9	Project document approved by parent group		

4 Notes
Background detail as needed.

5 Revision History
This section lists the dates the Work Plan and Schedule was modified.

1) Revised date

Template 8   Change Request

Project Description:			Project Start Date:	Project Manager
Each change description should cover a single modification. Attach supporting documentation as needed				
Change No.	Change Description	Originator	Resolution	Status
1				
2				
3				
4				
5				
6				
7				
8				
9				
10				
11				
12				
13				
14				
15				
16				
17				
18				
19				

9.2 Final Thoughts

Throughout this book we have stressed the importance of documentation and process. While it is easy to grumble about the time required to perform the tasks dictated by these steps, the effort is nonetheless essential to a modern organization. No one person has all the answers or all the great ideas. Process is the mechanism to bring together different ideas and fashion them into a strategy that accomplishes the task assigned by management. Documentation is the way ideas are communicated. Both work together to further the goals of the organization.

There is an old saying: "Anything worth doing is worth doing well." That's the core of documentation and process.

Index